T0213957

Introduction to Modeling and Numerical Methods for Biomedical and Chemical Engineers

Edward Gatzke

Introduction to Modeling and Numerical Methods for Biomedical and Chemical Engineers

 Springer

Edward Gatzke
Department of Chemical Engineering
University of South Carolina
Columbia, SC, USA

ISBN 978-3-030-76451-7   ISBN 978-3-030-76449-4   (eBook)
https://doi.org/10.1007/978-3-030-76449-4

This Springer imprint is published by the registered company Springer Nature Switzerland AG
The registered company address is: Gewerbestrasse 11, 6330 Cham, Switzerland

Dedicated to Andi, Drew, and Ellie.

Preface

This book is intended for chemical and biomedical engineering students in their first or second year of studies. Students should have at least completed a first course in differential calculus. Some topics are included for students that have completed vector calculus. Experience with differential equations and linear algebra is not assumed. Basic knowledge of physics is expected for topics related to forces. Additionally, some basic knowledge of chemistry is expected with relation to concentration and reaction rates.

This book was originally in support of an introductory course for biomedical engineers. The students mostly were in their second semester, having completed one course in calculus. The material was adopted to a course for chemical engineers who had completed the calculus sequence and were co-currently taking a class in differential equations. In both cases, a full course in linear algebra was not required, so this book introduces basic linear algebra topics. No computer programming course is required before the presented topics, although it certainly can be beneficial in many aspects of the course.

Engineering students face a daunting challenge: they must master advanced mathematics, they must gain scientific knowledge, and they should be able to solve technical problems combining math and science. In today's workplace, problem solving usually involves using computers. As a result, engineering students should develop the ability to solve engineering problems with computers. Some people are good at math and some people work well with computers. Engineers must use both math and computers while translating real-world situations into solvable problems.

Introductory engineering classes face many challenges. Students must start to learn how to translate engineering problems into a numerical representation using scientific fundamentals. There are engineering models that represent systems mathematically, there are numerical methods that can be used to solve the relevant equations, and there are software packages that help us implement the numerical methods to find the solution. The goal of this book is to provide resources related to all three topics: deriving mathematical engineering models, setting up numerical methods, and applying computers to solve for a solution.

Modeling methods considered include basic introductions to modeling of mechanical, electrical, and chemical systems. Included are the concepts of force/vector resolution, static equilibrium, Kirchoff's laws, mass balances, and materials models. A variety of numerical methods are presented at a very introductory level. The specific topics illustrated include matrix multiplication, matrix inversion, root finding, numerical integration, numerical optimization, and numerical solution of differential equations. To implement solution strategies, modern engineering students must become proficient using computer software packages. Software topics considered include spreadsheets (Excel), MATLAB, and COMSOL Multiphysics. Only a brief overview of the relevant software products are included. Laboratory exercises are available online so that they can be more rapidly updated for new software versions.

Columbia, SC, USA   Edward Gatzke
July 2020

Contents

Chapter 1
Modern Engineering

1.1 Engineering

Engineers solve problems using math and science. Why do students choose to suffer through so many difficult math, science, and engineering courses? Many are motivated by money. Engineering graduates are generally the highest paid college graduates based on starting salary. Companies are willing to give high salaries because they expect engineers to make money for the company. How do they make money? **Engineers make things better.** What are some examples of making things better? Engineers could:

- Design a stronger and lighter cycling helmet
- Create an improved bioethanol fermentor
- Devise a process to produce a superior personal body wash
- Invent a more efficient reactor system to make gasoline
- Develop software and hardware to easily track daily exercise.

You should notice two things about the above examples:

1. The verbs are all similar: *Design, Create, Devise, Invent, and Develop.*
2. The adjectives and adverbs are all similar: *Stronger, Lighter, Improved, Superior, Efficient, and Easily.*

How Do Engineers Make Things Better? They generally will not go to a workshop and tinker around until they eventually come up with something better. Engineers typically will use their math, science, and engineering background to analyze the problem. In many cases, this means setting up a problem mathematically and solving the resulting problem for a solution. The greatest skill an engineer can have is knowing what problem to solve.

There are many different strategies when working to solve a problem. One way is to draw a picture, a sketch, or a diagram. The sketch may not be 100% accurate,

© Springer Nature Switzerland AG 2022
E. Gatzke, *Introduction to Modeling and Numerical Methods for Biomedical and Chemical Engineers*, https://doi.org/10.1007/978-3-030-76449-4_1

but it should include the important information. Sometimes, it has extra information that will not be needed to solve the problem. There is a trade-off between having too little detail and having too much. Initial sketches may have limited information, while final design drawing may be very detailed with precise numerical values.

Engineers must make assumptions about the problem. What forces are acting on the device? What chemical reaction is taking place? Is the system changing with time? Is the process operating at a constant temperature? What materials are being used? What effects can be neglected to simplify the analysis? Engineers must learn either through experience or otherwise what assumptions are valid under what conditions.

Next, an engineer will determine the relevant model for the system. In many cases, this involves setting up mathematical equations to be solved. Can the ideal gas law be used for the gas in the reactor? What are the conditions for static equilibrium for the prosthesis? Which mechanical model best represents the stress–strain relationship for the material? What reaction rate expression should be used? The mathematical model helps the engineer deal with the complexity of a real system in a rational and logical manner. The model ideally serves a purpose in the process of making something better.

At this point, a solution is typically the next step. To do this, one must first understand both the fundamental mathematics and scientific principles. Mathematics is like the alphabet and science is like words. Engineering is about telling a story by pulling it all together to find a solution. Engineering is the combination of science and math for a purpose.

1.1.1 Chemical Engineering

Chemical engineers obviously deal with chemistry. What does this actually mean? Chemical engineers use chemical phenomena to accomplish a given task. This leads to using chemical reactions to convert raw materials into higher value substances. In many cases, the substances are not pure enough, and separation processes must be used to purify the materials. The fundamental principles of reaction and separation distinguish the chemical engineering discipline. Chemical engineers often think more about how to develop or improve a chemical process as opposed to many engineering disciplines that look to develop products. Chemical engineers work in many industries. Some industries include:

- Petrochemical: Petrochemical industries work to converting crude oil into usable products like gasoline and diesel fuel.
- Consumer Products: Companies specializing in consumer products make everything from house paint to laundry detergent to face cream.
- Paper Industry: The paper industry converts wood into paper and other wood products like paperboard and corrugated medium.

- Specialty Chemicals: Industries working in specialty chemicals create materials that are used by other industries such as lubricants and glues.
- Polymers: Polymer corporations make plastics and related polymeric materials, usually starting from organic feedstocks.
- Energy Production: Many chemical engineers work in energy production, using stored chemical energy to generate electricity.
- Environmental: Environmental engineering works in the area of monitoring and mitigating industrial environmental impact.
- Biotechnology: The biotechnology sector works with chemistry and biochemistry to produce drugs, biomaterials, and biomedical devices.

Obviously, these industries often overlap significantly; for example, petrochemical plant may also produce some polymers and specialty chemicals.

Many chemical engineers are considered "process engineers" because they work on a single production process in a very large site. A chemical process may often be summarized in a Process Flow Diagram (PFD). The PFD may summarize connections between various pieces of equipment used to react and separate materials. Usually, chemical plants involve many types of liquids, gases, and solids. The PFD typically shows how materials are processed in the facility.

All engineers must be aware of the potential impact of their work. Chemical engineers must be especially diligent, as chemical processes often contain extremely dangerous materials that could result in significant loss of life and environmental impact.

1.1.2 *Biomedical Engineering*

Biomedical engineers work on biomedical applications. In some cases, this could require designing an electronic device to monitor a patient. In other cases, mechanical devices could be required to help aid the healing process. Some biomedical engineers work on drug delivery systems. This means that biomedical engineers must understand mechanical, electrical, and chemical engineering fundamentals. They must also comprehend biology and biochemistry as it impacts medical devices and processes.

Some students choose to major in biomedical engineering because they assume that it will be less challenging than other engineering disciplines. This is not true. Biomedical engineers must master similar levels of mathematics. Biomedical engineers also encounter difficult topics common to other engineering disciplines such as fluid dynamics and thermodynamics. If anything, biomedical engineering is more difficult because of the numerous science courses and the wide variety of engineering topics to consider. Biomedical engineers must master the mechanics and dynamics of mechanical systems, the reaction and thermodynamics of chemical systems, and the electronics and instrumentation of electrical systems. Additionally, biomedical engineers are expected to understand material properties and how materials interact with biological systems.

Many biomedical engineering students will go on to medical school. Some engineers who went on to become practicing clinical doctors relate that medical school training was very different from their engineering training; rather than a focus on solving problems using knowledge and creativity, medical school courses focus mainly on memorization of relevant facts. Although often seen in medical shows on television, unique, innovative, and creative medical solutions are often left to biomedical research professionals rather than practicing physicians.

Like chemical engineers, biomedical engineers must understand the potential impact of their work. While an inept doctor may maim or kill a few people before being discovered, a poorly designed medical device has the potential to impact hundreds or thousands of lives. This ethical consideration is especially challenging due to the complexity of the human body. The broader impacts in the medical industry may often be difficult to reliably determine.

1.1.3   Other Engineering Disciplines

Mechanical Engineering Mechanical engineers work on design and production of mechanical systems. They may be employed at discrete manufacturing plants making everything from cars to dishwashers. Producing a product can be more challenging than designing a product.

Civil Engineering Civil engineers design the built environment we live in. This includes everything from buildings to roads. They work on other infrastructure, including our transportation infrastructure.

Electrical Engineering Electrical engineers work to design products based on electrical properties and phenomena. They work on everything from small consumer devices to complex computers and the electric power distribution grid.

Industrial Engineering Industrial engineers work to improve industrial practices. They consider efficiency improvements ranging from factory production to warehouse design or even the entire product distribution network.

Computer Engineering Computer engineers are often seen as a combination of electrical engineering and computer science, specializing in the design and production of computer system hardware.

Aerospace Engineering Aerospace engineers work on designing and building aeronautical and space applications.

Nuclear Engineering Nuclear engineers work on the safe use of nuclear reactions, typically for power production.

Materials Engineering Materials engineers study and use properties of materials, like alloys, composites, ceramics, and polymers.

Petroleum Engineering Petroleum engineers work in areas related to oil discovery and drilling.

1.2 General Problem Solving Strategies

Engineers work to solve problems. There are many strategies for approaching engineering problems. One approach is the Dartmouth/Thayer method. This process is summarized as follows:

1. Problem Statement: What is the goal?
2. Redefine the Problem: Can the objective be simplified?
3. Develop General Specifications: Define constraints and limitations.
4. Brainstorm Alternatives: What different solutions are there?
5. Select Best Alternatives: Formulate a selection criterion.
6. Check Problem Definition: Make sure that the alternative solves the problem.
7. Redefine and Add Specifications: Modify and add limitations if needed.
8. Brainstorm Again: Consider alternatives again.
9. Reiterate Until Adequately Solved: Keep improving until good enough.

The engineering design method is comparable to the scientific method. However, they have different objectives. The scientific method works to discover the truth about our world. The engineering design method works to improve our world. Engineering problems have an objective or design criterion. The resulting solution may need to be lightweight, strong, or low-cost. The solution may be constrained with engineering specifications. These limits dictate what is feasible for the solution, such as the minimum strength or the maximum cost (Fig. 1.1).

1.2.1 Sketches and Schematics

Drawing pictures is an important part of engineering. Each engineering domain has certain types of typical schematics. For example, an electrical engineer may draw a simple circuit. A mechanical engineer may draw a free body diagram indicating the relevant forces. Chemical engineers generally work more with a process, rather than a product. The process diagram would include the relevant steps in the production process. This is seen in Fig. 1.2, a simple fermentation process to make ethanol.

This sketch shows the overall process. The system exists at many levels of detail. Chemical engineers could work on designing a component in a process. In the fermentation process, the reactor and distillation process steps could be considered individually. There are various types of reactors (batch, semibatch, continuous

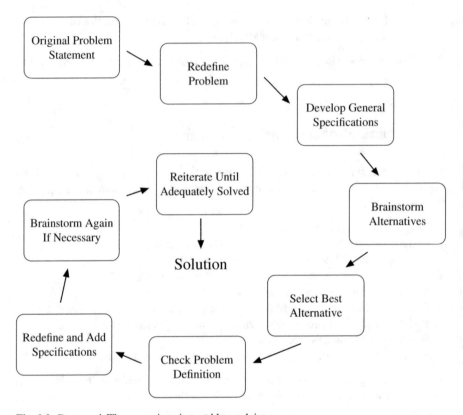

Fig. 1.1 Dartmouth/Thayer engineering problem solving

stirred tank, etc.) that could be used, as there are many types of separation methods that could be considered. At a lower level, a chemical engineer could examine in detail a portion of a single unit. In a distillation process, liquid and vapor phases interact and flow together and are impacted by the physical design of the individual "trays" that make up the distillation tower. At even a smaller level, the underlying molecular interactions lead to the chemical properties that drive the design of the process. The reaction pathway itself is the fundamental core of the process. If sugar cannot be converted to ethanol, there is no reason to consider the process. Understanding the chemical reactions can lend insight into improving and designing the process.

In a similar way, biomedical engineers must understand the biochemical reactions and phenomena. In some cases, biomedical engineers may work in a production environment where they must use multiple steps to produce the desired product; for example, pharmaceutical production may involve working with very specialized production processes such as protein extraction or biological reactors. In other cases, biomedical engineers may work on mechanical or electrical devices to solve specific

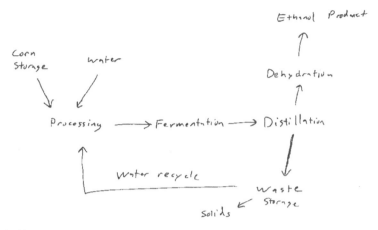

Fig. 1.2 Simple hand-drawn process flow diagram for a corn fermentation and distillation process. A detailed schematic is often not required when initially considering modeling a process

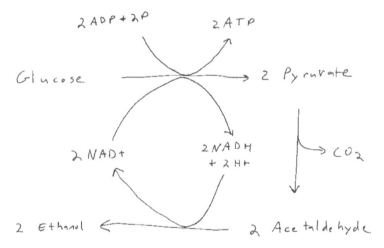

Fig. 1.3 Simple hand-drawn fermentation process describing biochemical reaction pathways relevant to glycolysis of glucose for ethanol production. Reaction rates and other specific information are not included in this type of sketch

medical problems. Whatever the task may be, it is often a good idea to draw a picture to help bring order to the chaos or complexity (Fig. 1.3).

1.2.2   Models and Numerical Solutions

Engineers can use scientific knowledge to describe their problems. In many cases, this allows for a numerical representation using relevant equations. The following

Fig. 1.4 A basic circuit
example

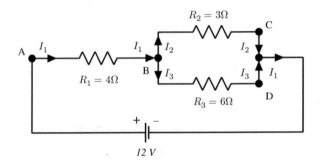

example considers using engineering principles to model a circuit system. Electrical
systems are considered in more detail in Sect. 4.2.

Electrical engineers must understand Kirchhoff's laws stating that the sum of
currents meeting at a point is zero and the sum of voltages in a loop is zero.
Mathematically, these two concepts can be represented in general form as:

$$\sum_{j=1}^{m} I_j = 0$$

$$\sum_{k=1}^{n} V_k = 0$$

Subscripts indicate that there could be multiple current and voltage values. Ohm's
law relates voltage, current, and resistance. Mathematically, this is written in the
general sense as:

$$V = IR$$

These general laws can be used together to solve for unknown values in circuit
problems. Writing out the governing equations allows for mathematical analysis.
Given enough information for unknown values, you can solve the mathematical
relationships for the numerical values of unknown quantities.

Consider the circuit in Fig. 1.4. Three current variables are defined: I_1, I_2, and I_3.
Three resistors are shown with numerical values for their actual resistance values:
$R_1 = 4\Omega$, $R_2 = 3\Omega$, and $R_3 = 6\Omega$. The voltage drop across each resistor may be
expressed using Ohm's law as:

$$V_{AB} = I_1 R_1$$

$$V_{BC} = I_2 R_2$$

$$V_{BD} = I_3 R_3$$

The electric potential at points C and D must be equal because they are connected to each other, so you may assume:

$$V_{BC} = V_{BD}$$

Kirchhoff's laws for the system can relate current at a branching point and voltage around the loop as:

$$I_1 - I_2 - I_3 = 0$$

$$12 - V_{AB} - V_{BC} = 0$$

There are now six algebraic relationships involving nine variables. Three of the variables explicitly defined in the problem: the numerical values of the resistances. The unknown values could now be calculated from these equations to determine the numerical values for unknown currents and voltages.

This is a relatively simple example that demonstrates how problems can quickly become complicated. This example problem can be solved by hand, but most engineering problems will require help from computers because the number of variables and equations can quickly grow. Additionally, many engineering relationships are complicated nonlinear expressions that cannot easily be solved using basic algebra. This book tries to help you grow beyond the small problems you can solve on your own to larger problems that you can set up and solve using computers and software.

In general, what strategies can be used to arrive at a solution? Sketches, diagrams, and schematics can help with visual interpretation of the system under consideration. These can help you define variables and describe what is happening in the physical real-world system. However, visual representations generally only help to determine what the mathematical representation should be. They are an aid in setting up the equations.

Engineering students should learn habits of writing clearly and neatly so that any derivation can be understood by others. A detailed sketch may be needed to illustrate required assumptions. The relevant equations should be written out for the problem. Known constant values should be written down, ideally with units. The unknown values can then be determined if adequate information and a proper formulation are provided for a well-posed problem.

1.3   Process Flow Charts

Engineers solve problems. You start at **Point A,** and then you want to get to **Point B**. The process of getting from A to B can be more formally considered in a process flow chart. The process flow chart can be for any activity, from filling orders to analyzing a data set to remove erroneous data points. Flow charts are ways to express a logical process in a visual manner. As seen in Fig. 1.5, there are standard shapes used for different actions in a flow chart. The oval shapes denote the starting and ending points. A trapezoid is used when data is either required or displayed. This is

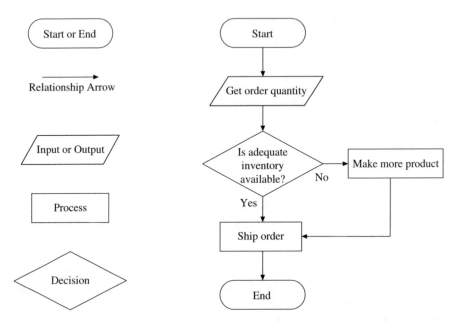

Fig. 1.5 Flow chart symbols for defining a process. Example showing a simplified order fulfill-ment process in a warehouse

often called process input or process output. Rectangular boxes are used to express a step in the process. Diamonds indicate a branch in the process where a condition is checked; the process continues on a route dependent on the condition value. The entire flow chart is tied together with directional arrows. When executing the process flow chart, each step proceeds sequentially after the current step has completed. Each step may change the state of the process. Each step should help along the path from A to B.

A process flow chart is useful in that it can help with visualization of the logic required to solve the problem. Expressing the same logic in a paragraph of text is obviously possible, but the mind may have difficulty in understanding the overall process. A visual flow chart representation helps present the systematic process with less ambiguity. Additionally, the sequential procedure for complex branched processes is easier to follow in a visual format.

The level of detail is important to aid in comprehension. A complex flow chart involving many different procedural routes can be difficult to grasp. Generally, a hierarchical approach is needed where a single step in a process may represent many actual steps. The example in Fig. 1.5 shows a single step for "Make more product," but this step obviously involves many individual steps at a finer level of detail.

A process flow chart should be a general solution to the problem, given assumptions about the problem. Consider making a flow chart to solve a simple maze. If you assume that the maze is always the same, the flow chart solution could be a list of basic moves. This solution is not general, in that it only works for one specific maze. Assuming that the maze is not known, a more general solution should be able to solve all cases. One possible solution would be "walk forward always keeping a wall on your right." This would be a more general solution as it can solve many mazes. However, additional assumptions should be made as some maze classes would make the solution fail.

1.4 Computers in Engineering

Computers are very good at performing repetitive calculations. Decades ago, engineers would work out mathematical problems by hand using only a pencil and mathematical tables. Years ago, engineers would work out mathematical problems using simple calculators. Current technology involves using general-purpose computers to solve engineering problems.

- Input: Any information that is used by the computer. Information may come from many sources such as the keyboard, mouse input, a microphone, an attached camera, a stored file, or even the Internet.
- Output: Any information that is produced by the computer. The information can be stored in a file, displayed on a screen, sent to a web site, or produced with audio speakers.
- Network: The modern information age was enabled by computer network connectivity. Information can be shared across the network, which is typically a TCP/IP packet-based network. Other network options include cellular data networks and Bluetooth wireless networks.
- Storage: Nonvolatile storage allows for information to be retained when the computer is turned off. This currently would typically be a hard disk drive or a solid-state drive. Volatile storage is the temporary memory that holds values only when the computer is on.
- Processing: The central processing unit (CPU) performs the calculations and manipulation of data coming from/going to input, output, network, or storage sources. A CPU may consist of millions of individual transistors. A modern CPU may have multiple cores. Each core is a separate CPU that shares memory and hardware access so that jobs can be performed in parallel.

To work effectively with computers, engineers should understand basic computational terms. Not everyone will be a computer scientist or computer engineer. Knowing how these powerful machines work will allow engineers to effectively and rapidly solve problems.

Engineers Must Know How to Work with Large Data Sets The data may have thousands of values and contain values corrupted by noise. Rather than sort through individual data points, simple computer programs can perform repetitive tasks like discarding bad values or calculating the average of the data set.

Engineers Must Know How to Visualize Large Data Sets In some cases, the problem or experiment may provide many data points. An engineer may not need to solve a specific problem but rather may need to determine trends in the data. With advanced plotting and visualization computational tools, engineers can quickly narrow down a problem.

Engineers Must Know How to Solve Big Problems In simple cases, a problem could be modeled as a handful of equations with a few unknowns. This type of toy problem could be solved in a quiz or exam to demonstrate knowledge of the class of problem. In reality, the relevant problem may involve writing and solving thousands of simultaneous equations. Computer systems allow engineers to solve problems that would be impossible using traditional methods.

Engineers Must Know How to Solve Hard Problems In many cases, an engineering problem may be best represented with a complex nonlinear equation. Basic mathematics handles simple linear situations. However, numerical methods can be used on a computer to determine an approximate numerical solution to a problem that cannot be easily solved by analytic methods.

Engineers Must Know How to Use Computers to Solve Problems Without computers, engineers would live in a world of slide rules and mechanical calculators. Computers allow complex problems to be solved in a reasonable amount of time with a small amount of human effort.

1.5   Suggestions for Homework Submissions

Engineering courses will assign homework problems. Problems can be quite complex, requiring multiple pages of work to find a solution. Below are a few suggestions that may help you be more successful when turning in problem sets. Every instructor will have their own requirements which are usually published in the course syllabus. A suggested general approach for assignments involving complex engineering problems is as follows:

1. Write out a brief summary of the problem statement.
2. Sketch or draw a diagram related to the problem as needed.
3. Describe any assumptions you are making.
4. Write out the equations that you are using to solve the problem.
5. Write down values for constants.
6. Label units for values and variables where appropriate.
7. Replace variables with numbers where appropriate.
8. Add notations to explain how you solved the problem.
9. Box your final answer, and remember to include units.
10. Write clearly, avoid tiny handwriting, and be neat.
11. Never write on the back of a piece of paper.
12. Start each major problem on a new piece of paper.

13. Never write your solution on the assignment handout unless told to do so.
14. Never tear out paper from a spiral bound notebook to turn in.
15. Use a cover sheet with your name, class, due date, and assignment number.
16. Always staple your assignment before turning it in.
17. Work in groups checking solutions and helping with methodologies but do not copy homework verbatim. Do not allow others to copy your homework.
18. You may have to work through the problem multiple times to get to a solution. If so, rewrite the solution in a clear and organized manner.
19. Print out graphs with legible axes labels and axes legends.
20. When turning in assignments electronically, use a PDF scanner to correct for white balance and parallax.

Problems

Use the blocks described in Fig. 1.5 to a possible flow chart for the following scenarios. Each block should represent one step in the overall process. Mentally trace your flow chart and consider the possible "what-if" scenarios.

1.1 Draw a flow chart for making a peanut butter sandwich. Assume that the peanut butter jar and the jelly jar may or may not be open. Use statements in your blocks like "Is the jelly jar open?" and "Spread peanut butter on one slice of bread." You may also ask questions: "Do you want the crust removed?"

1.2 Given a large jar full of various coins, draw a flow chart for sorting the coins into separate containers, assuming you can only remove one coin at a time. You may check to see if the jar is empty.

1.3 Given a jar full of one type of coins, draw a flow chart for stacking the coins in piles of 10. You may check to see if the jar is empty.

1.4 Use a variable named *count*. Given a jar full of various coins, draw a flow chart to determine the total number of one type of coins in the jar. Assume that you can only remove one coin at a time. Remember to set the initial value of the variable *count*. You may check to see if the jar is empty.

1.5 Use a variable named *total*. Given a jar full of various coins, draw a flow chart to determine the total value of the jar. Assume that you can only remove one coin at a time. Remember to set the initial value of the variable *total*. You may check to see if the jar is empty.

1.6 Use a variable named *vmax*. Given a jar full of balls labeled with positive integer values, draw a flow chart to determine the maximum value of any one ball in the jar. Assume that you can only remove one ball at a time. You may check to see if the jar is empty. Hint: if you initially set the variable *vmax* to 0, the first ball you check will become the maximum value found so far in your search. That value will be replaced when you find a ball with a larger value.

1.7 Use variables named **rcount** and **gcount**. Given a reading assignment, draw a flow chart to determine the total number of times the word "red" appears in the text and also how many times the word "green" appears in the text. Assume that you can only read one word at a time. Remember to set the initial value of the variables. You may check to see if there are any words left in the text.

1.8 Use variables named *fact* and **value**. Draw a flow chart to compute the factorial for an input value and store it in the variable *fact*. Ask for a positive integer as input. Assign the input to variable **value**. Include error checking to see if the input value is a positive integer before proceeding.

1.9 Use variables named **eavg** and **count**. Given a list of positive integers, draw a flow chart to determine the average of the even values. You may read one number at a time, you may check to see if it is even, and you may check to see if there are any more values in the list. Hint: sum up the even values in the variable **eavg** while also keeping a count of how many even values you find.

Further Reading

1. Attaway, S. (2018). *MATLAB: A practical introduction to programming and problem solving* (5th ed.). Butterworth-Heinemann.
2. Chaudhiri, A.B. (2005). *The art of programming through flowcharts and algorithms*. Laxmi Publications.
3. Chaudhiri, A.B. (2020). *Flowchart and algorithm basics: The art of programming*. Mercury Learning and Information.
4. Eide, A., Jenison, R., Northup, L., & Mickelson, S. (2017). *Engineering fundamentals and problem solving* (7th ed.). McGraw-Hill Education.
5. Farrell, J. (1994). *Computer programming logic using flowcharts*. Boyd and Fraser Pub Co.
6. Fogler, H.S., LeBlanc, S.E., & Rizzo, B. (2013). *Strategies for creative problem solving* (3rd ed.). Pearson.

Chapter 2
Mathematical Fundamentals

Unlike other chapters, this chapter can be treated as a single assignment. Students should read through this chapter and complete all the problems that are labeled in the form (P2.xx), where xx is the problem number. Try to dedicate substantial effort to complete this assignment as this material is fundamental to other later topics. Sections 2.8 and 2.9 assume a basic knowledge of derivative calculus.

2.1 Basic Math

Engineers must know basic mathematical operations. Mainly, this means understanding the order of operations and how to use a calculator.

> **Order of Operations**: Mathematical expressions are evaluated in a defined order. The initialism commonly used to express the order of operations is **PEMDAS**. This means that one should evaluate expressions in parentheses first, then exponential operations, next multiplication and division, and finally addition and subtraction.

Problems: Basic Math

2.1 Evaluate the following and simplify your answer.

$$\frac{(5+3)^2}{4}$$

© Springer Nature Switzerland AG 2022
E. Gatzke, *Introduction to Modeling and Numerical Methods for Biomedical and Chemical Engineers*, https://doi.org/10.1007/978-3-030-76449-4_2

2.2 Evaluate the following and simplify your answer.

$$\left(\frac{2}{32+48}\right)^{-1}$$

2.3 Evaluate the following and simplify your answer.

$$\cos(4\pi)$$

2.4 Evaluate the following and simplify your answer.

$$\sqrt[3]{150-25}$$

2.5 Evaluate the following and simplify your answer.

$$\sin\left(\pi+\frac{\pi}{2}\right)$$

2.6 Evaluate the following and simplify your answer.

$$2+3^2-(2-3)$$

2.7 Evaluate the following and simplify your answer.

$$(1000/100)^2$$

2.8 Evaluate the following and simplify your answer.

$$\frac{x^2-3x}{2x-6},\quad x=14$$

2.9 Evaluate the following and simplify your answer.

$$\frac{4-2}{1+9}+\frac{3}{3+6}$$

2.10 Evaluate the following and simplify your answer.

$$\frac{1}{1+\frac{2}{6+2}}$$

2.2 Basic Algebra

Students should know how to use basic algebraic properties such as the commutative, associative, and distributive properties. As a reminder, commutative means $ab = ba$, associative means $(ab)c = a(bc)$, and distributive means $ac + bc = (a + b)c$. You should know how to simplify expressions by canceling terms, combing similar terms, and removing common factors.

Engineering students will have to solve equations that include a variety of constants, not just numbers. Many engineering equations will be written in terms of numerous variables, not just simple numbers and one or two variables. Consider solving for x in the following equation:

$$ax = by + cx$$

with constants a, b, and c along with variable y.

- First, get all the terms with x on one side by subtracting cx from both sides: $ax - cx = by$.

  - Remember, a **term** in an equation is a group of expressions added or subtracted from the equation.

- Next, use the distributive property to "pull out" the variable x from the two terms: $(a - c)x = by$.
- Finally divide both sides by $(a - c)$ to solve for x in terms of y and some constants: $x = \frac{by}{(a-c)}$.

  - Note: this assumes that $a - c$ is not equal to zero.

Now, the whole sequence of steps can be shown as:

$$ax = by + cx$$
$$ax - cx = by$$
$$(a - c)x = by$$
$$x = \frac{by}{(a - c)} = \left(\frac{b}{a - c}\right) y$$

Roots of Equations

If you have a function of one variable, you may be able to find a *solution* to the equation. This means that you find a value of x that *satisfies* the equation. The values of x that satisfy the equation are also called the *roots* of the equation. In the general form, this is often written as:

$$f(x) = 0$$

In some cases, one may easily solve the equation *analytically*. This means that you get a closed-form expression for the solution that satisfies the equation. The closed-form expression is not an approximation.

Example 2.1 (Roots of a Simple Function) Consider the function $f(x) = x^2 - 2$. This is a parabola with a minimum value at $x = 0$ where $f(x) = -2$. The values of x that satisfy the equation are $x = \sqrt{2}$ and $x = -\sqrt{2}$. The value $\sqrt{2}$ is an exact analytical expression that can be approximated numerically as 1.41421356 with some truncation error. Here, error is less than 10^{-8}.

Quadratic Equation

For the quadratic equation, $ax^2 + bx + c$, the roots are $x = \frac{-b \pm \sqrt{b^2 - 4ac}}{2a}$. Note that imaginary roots do not always mean that something is incorrect. In many process systems engineering problems, roots of a polynomial should have imaginary components.

Repeated Roots

For the function $f(x) = x^3$, $x = 0$ is the solution to the equation. Actually, there are three roots, all of them occurring at a value of $x = 0$. Again, multiple roots to an equation can have significant meaning in some applications.

Polynomials

In many cases, you may have a simple polynomial function that requires the roots to be found. A simple polynomial can be expressed as:

$$4x^3 + 2x^2 - 7x + 1 = 0$$

A more general form could be:

$$a_n x^n + a_{n-1} x^{n-1} + \ldots a_2 x^2 + a_1 x + a_0 = 0$$

As in the previous example, this is a polynomial function. In this case, the polynomial is a function of the variable x with multiple constants a_i.

- In the general case, $(x - r_1)(x - r_2) \ldots (x - r_n) = 0$, roots $= r_1, r_2, \ldots r_n$ will satisfy $f(x) = 0$.

  - Remember, sometimes roots can be repeated. This can have an impact on solutions in some cases.

- In the specific second-order case, the equation is: $ax^2 + bx + c = 0$. This quadratic equation has two roots at $x = \frac{-b \pm \sqrt{b^2 - 4ac}}{2a}$.
- There are analytical expressions for roots of polynomials up to fifth order. However, exact higher-order solutions are very complex.

Problems: Finding Solutions

2.11 Find all solutions that satisfy the following algebraic relationship:

$$x\,(x-1)\,(x+3) = 0$$

2.12 Find all solutions that satisfy the following algebraic relationship:

$$x^2 + x = 2$$

2.13 Find all solutions that satisfy the following algebraic relationship:

$$x^3 + x = 0$$

2.14 Find all solutions that satisfy the following algebraic relationship:

$$2x^2 + 3x + 1 = 0$$

2.15 Find all solutions that satisfy the following algebraic relationship:

$$-10x - 28 = -2x^2$$

2.16 Find all solutions that satisfy the following algebraic relationship:

$$x^2 + 5x + 6 = 0$$

2.17 Find all solutions that satisfy the following algebraic relationship:

$$\frac{1}{x} = \frac{x}{x+6}$$

2.18 Find all solutions that satisfy the following algebraic relationship:

$$\frac{(x-3)+1}{1+3} = \frac{1}{4}$$

2.19 Find all solutions that satisfy the following algebraic relationship:

$$x^2 - 2x - 3 = 0$$

2.20 Find all solutions that satisfy the following algebraic relationship:

$$3x + \frac{8}{3+1} = 2 - x^2$$

2.3   Sets of Linear Equations

You can solve sets of linear equations. When you have n equations and n unknowns, a simple solution methodology is to use the first equation and solve for a variable. In the next equation, replace that variable with values you found from the first equation and solve for a different variable. Keep eliminating variables until you find a solution. This procedure may be formalized for linear systems using row reduction methods or more general linear algebra formulations.

Problems: Linear Equations

2.21 Find the solution to the following set of equations:

$$14 = x + 2y$$
$$17 = 3x + y$$
$$1 = z$$

2.22 Find the solution to the following set of equations:

$$3 = x$$
$$7 = -2y - z$$
$$-14 = y + 4z$$

2.23 Find the solution to the following set of equations:

$$7 = x + 0y$$
$$0 = x + y + z$$
$$-1 = 0x + z$$

2.24 Find the solution to the following set of equations:

$$16 = x + z$$
$$-7 = 2y + z$$
$$-15 = -x + y$$

2.25 Find the solution to the following set of equations:

$$-1 = y + z$$

$$2 = -x - z$$
$$-2 = y + 2z$$

2.26 Find the solution to the following set of equations:

$$-1 = 0x + 1y + 1z$$
$$2 = -1x + 0y - 1z$$
$$-2 = 0x + 1y + 2z$$

2.27 Find the solution to the following set of equations:

$$4 = 1x + 0y + 1z$$
$$0 = 1x + 1y + 0z$$
$$0 = 0x + 1y + 1z$$

2.28 Find the solution to the following set of equations:

$$7 = -3x - y + z$$
$$0 = x + 2y$$
$$7 = -y + z$$

2.29 Find the solution to the following set of equations:

$$0 = 17x + 2y + 4z$$
$$0 = 15x + 9y - 8z$$
$$0 = -3x + 8y - 4z$$

2.30 Find the solution to the following set of equations:

$$16 = x + z$$
$$17 = x + y$$
$$8 = x - y + z$$

2.4   Unit Conversion

Engineers must know how to convert units. Additionally, engineers should recognize common metric prefixes such as kilo-, centi-, milli-, and micro- along with their abbreviations; for example, the relevant units of length are km, cm, mm, and μm. To convert from one type of units to another, just multiply by the conversion factor and cancel the relevant units. This can also be done for more complex units such as area, volume, concentration, and others.

Example 2.2 (Unit Conversion) To convert units, just multiply by the conversion factor and cancel units. To convert 2.5 ft to inches, you must know the conversion factor that relates the two units.

$$2.5 \ ft = ?? \ inches$$

$$2.5 \ ft \quad \frac{12 \ inches}{1 \ ft} = ?? \ inches$$

$$2.5 \ \cancel{ft} \quad \frac{12 \ inches}{1 \ \cancel{ft}} = 30 \ inches$$

$$2.5 \ ft = 30 \ inches$$

Note that some engineering problems include non-metric units. A few useful English unit conversions include the following:

- 1 foot is 30.48 cm.
  - A one-foot ruler is about 30 cm.
- 1 gallon is 3.79 L.
  - A fifth of whiskey is about 1/5 gallon.
- 1 kg is 2.2 pounds.

Additionally, temperature conversion can be very important. Celsius, Fahrenheit, Kelvin, and Rankine are all used in engineering calculations. If you remember 0 °C is 32 °F and 100 °C is 212 °F, conversion formulas can easily be derived such that:

$$F = \left(C \ \frac{9}{5} \right) + 32$$

$$C = (F - 32) \frac{5}{9}$$

Often, significant effort is required to analyze the units in a problem.

Problems: Unit Conversion

Find numerical values for the following. Remember to include units in your answers.

2.31 Convert 650 cm^3 to L.

2.32 Convert 0.013 L to cm^3

2.33 Convert 52,100 kg to g.

2.34 Convert 9.2 mg to kg.

2.35 Convert 2.3 m^2 to cm^2.

2.36 Convert 6340 cm^2 to m^2.

2.37 Convert 4300 μg to mg.

2.38 Convert 2.3 gal to L.

2.39 Convert 42 °C to °F.

2.40 Convert 178 lb to kg.

2.5 Reading Graphs

Often, data is provided in a figure or a graph. Engineers must be able to read a wide variety of graphs to determine numerical values. A simple two-dimensional plot includes the abscissa (horizontal x direction) and ordinate (vertical y direction). Engineering plots generally relate to real data, so a value of interest may be plotted as a function of some other variable. For example, a flow rate value could be plotted as a function of time. Reaction rate values could be graphed as a function of temperature. Engineers must be able to move between x–y mathematical notation and real-world values.

Typically, single data values are plotted on a graph with a single dot or icon representing each point. The data points are not generally joined by a solid line. However, the data can lead to development of a model. The model may be a function that has a value at each horizontal point on the graph.

Engineers must also be able to find the slope (derivative) of a function. For a function in the form $y = f(x)$, this means finding $\frac{df}{dx}(x)$, which is also a function of x. For a function of time, $y = f(t)$ and the rate of change is the slope of $f(t)$. The rate of change means finding $\frac{df}{dt}(t)$.

Fig. 2.1 First example graph

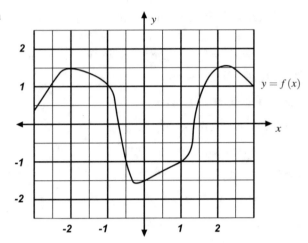

Fig. 2.2 Second example graph

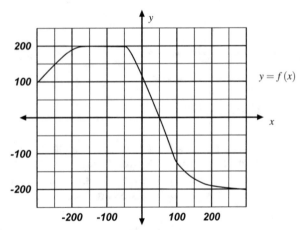

Problems: Graphs of Functions

2.41 Estimate $f(x)$ at $x = -1$ from Fig. 2.1.

2.42 Estimate $\frac{df}{dx}$ at $x = -2$ from Fig. 2.1.

2.43 Estimate $\frac{df}{dx}$ at $x = 0$ from Fig. 2.1.

2.44 Estimate $\frac{df}{dx}$ at $x = 3$ from Fig. 2.1.

2.45 Estimate $f(x)$ at $x = 3$ from Fig. 2.1.

2.46 Estimate $f(x)$ at $x = -100$ from Fig. 2.2.

2.47 Estimate $\frac{df}{dx}$ at $x = -100$ from Fig. 2.2.

2.48 Estimate $\frac{df}{dx}$ at $x = 0$ from Fig. 2.2.

2.49 Estimate $f(x)$ at $x = 50$ from Fig. 2.2.

2.50 Estimate $f(x)$ at $x = -30$ from Fig. 2.2.

2.6 Making Graphs

You should know how to graph functions without the use of a calculator or a computer. For some functions, you may want to pick a variety of values of the independent variable x and evaluate the function values, and then graph $y = f(x)$ *vs.* x. Typically, x represents the independent variable and y represents the dependent variable.

> **Dependant and Independent Variables**: A mathematical function may be used to define the value of a dependent variable which is dictated by the value of an independent variable. The independent variable may be considered the input to the function and the dependent variable may be considered the output of the function.

For a sum of two functions $f = f_1 + f_2$, you can plot f_1 and f_2 and add the functions point by point. In many cases, you need to only graph the "interesting" points of the response where something significantly changes. Interesting points could be at $x = 0$, $x = 1$, or $x = \infty$. For trigonometric functions, multiples of $\pi/2$ may be "interesting" values.

Problems: Sketching Functions

2.51 Sketch the graph of the following function:

$$y = 2x + 1$$

2.52 Sketch the graph of the following function:

$$y = \frac{1}{x}$$

2.53 Sketch the graph of the following function:

$$y = \sin(x)$$

2.54 Sketch the graph of the following function:

$$y = x^2 + 1$$

2.55 Sketch the graph of the following function:

$$y = 2\sqrt{x}$$

2.56 Sketch the graph of the following function:

$$(x - 1)^2 + (y - 1)^2 = 4$$

2.57 Sketch the graph of the following function:

$$x^2 = y + 2$$

2.58 Sketch the graph of the following function:

$$y = \frac{1}{x} - 1$$

2.59 Sketch the graph of the following function:

$$y = e^x$$

2.60 Sketch the graph of the following function:

$$y = x$$

2.7　Trigonometry

Trigonometry relates to solving problems with triangles. Engineers often deal with forces acting on a solid, liquid, or gas. Forces can be broken into Cartesian components simplifying solution of a given problem. This requires using trigonometric concepts to solve the problems.

Problems: Graphs of Functions

Express the answer to the following questions in terms of a, b, c, and θ. You may need to use trigonometric functions such as sin, cos, and tan. Also, remember that

Fig. 2.3 Right angle
triangles for trig function
problems 2.61–2.70

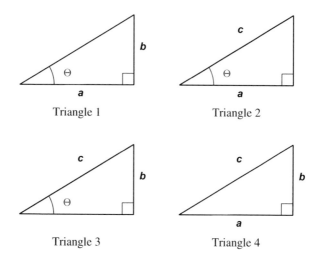

Triangle 1   Triangle 2

Triangle 3   Triangle 4

the Pythagorean Theorem holds for right triangles. Do not numerically evaluate the
answer. For each triangle, assume that you can only use the variables shown.

2.61 Find an expression for θ from Triangle 1 in Fig. 2.3.

2.62 Find an expression for a from Triangle 1 in Fig. 2.3.

2.63 Find an expression for b from Triangle 1 in Fig. 2.3.

2.64 Find an expression for θ from Triangle 2 in Fig. 2.3.

2.65 Find an expression for a from Triangle 2 in Fig. 2.3.

2.66 Find an expression for c from Triangle 2 in Fig. 2.3.

2.67 Find an expression for θ from Triangle 3 in Fig. 2.3.

2.68 Find an expression for a from Triangle 4 in Fig. 2.3.

2.69 Find an expression for b from Triangle 4 in Fig. 2.3.

2.70 Find an expression for c from Triangle 4 in Fig. 2.3.

2.71 Find an expression for θ from Triangle 1 in Fig. 2.4.

2.72 Find an expression for a from Triangle 1 in Fig. 2.4.

2.73 Find an expression for b from Triangle 1 in Fig. 2.4.

2.74 Find an expression for θ from Triangle 2 in Fig. 2.4.

2.75 Find an expression for a from Triangle 2 in Fig. 2.4.

2.76 Find an expression for c from Triangle 2 in Fig. 2.4.

2.77 Find an expression for θ from Triangle 3 in Fig. 2.4.

Fig. 2.4 Right angle
triangles for trig function
problems 2.71–2.80

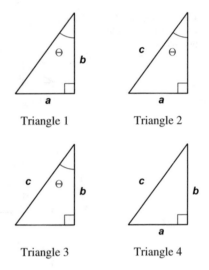

Triangle 1   Triangle 2

Triangle 3   Triangle 4

2.78 Find an expression for a from Triangle 4 in Fig. 2.4.

2.79 Find an expression for b from Triangle 4 in Fig. 2.4.

2.80 Find an expression for c from Triangle 4 in Fig. 2.4.

2.8   Calculus: Basic Derivatives

Engineers must understand the concept of a derivative function. If you are given the
function, you should know the derivative of simple functions such as $3x^4$, $\frac{1}{x}$, $2e^{-3t}$,
and $\sin(2t)$. You should also remember how to use the chain rule.

Derivative Function: The derivative function is the function that describes the
slope of a given function.

Just as a function can be evaluated at a given value, a derivative can also be
evaluated at a given value. Remember that the derivative evaluated at a point,
$\frac{df}{dt}(t)\big|_{t=t^*}$, is the slope of the function $f(t)$ at time $t = t^*$. This also defines the
slope of the line tangent to $f(t)$ at time $t = t^*$.

Problems: Basic Derivatives

2.81 Evaluate $\frac{df}{dx}$ at $x = 0$ for

$$f(x) = 2 + 3x + 4x^2$$

2.82 Evaluate $\frac{df}{dx}$ at $x = -1$ for

$$f(x) = x^{-2}$$

2.83 Evaluate $\frac{df}{dx}$ at $x = 0$ for

$$f(x) = 3\cos(x)$$

2.84 Evaluate $\frac{df}{dx}$ at $x = 3$ for

$$f(x) = 2x + 6x$$

2.85 Evaluate $\frac{df}{dx}$ at $x = 0$ for

$$f(x) = e^x$$

2.86 Evaluate $\frac{df}{dx}$ at $x = 1$ for

$$f(x) = \frac{1}{x}$$

2.87 Evaluate $\frac{df}{dx}$ at $x = 4$ for

$$f(x) = \sqrt{x}$$

2.88 Evaluate $\frac{df}{dx}$ at $x = 0$ for

$$f(x) = 2\sin(x)$$

2.89 Evaluate $\frac{df}{dx}$ at $x = 1$ for

$$f(x) = x^5$$

2.90 Evaluate $\frac{df}{dx}$ at $x = 5$ for

$$f(x) = 3 + \frac{1}{4}$$

2.9   Calculus: Advanced Derivatives

You should be familiar with the chain rule for more complex functions. Basically, given a complex function,

$$F(x) = f(g(x))$$

The derivative of $F(x)$ is represented as:

$$\frac{dF}{dx} = \frac{df}{dg}\frac{dg}{dx}$$

or using Lagrange notation, the first derivative is given as F':

$$F'(x) = f'(g(x))\,g'(x)$$

Example 2.3 (Chain Rule) To determine the derivative of the function:

$$F(x) = 2e^{-3x}$$

first realize that this can be expressed as the function of a function:

$$F(x) = 2e^{g(x)} = f(g(x))$$

where $g(x) = -3x$ and $f(x) = 2e^x$. You know that $g'(x) = -3$ and $f'(x) = 2e^x$. Therefore, $f'(g(x)) = 2e^{-3x}$ and:

$$F'(x) = f'(g(x))\,g'(x) = 2e^{-3x}\,(-3) = -6e^{3x}$$

Problems: Advanced Derivatives

2.91 Evaluate $\frac{df}{dx}$ at $x = 0$ for

$$f(x) = e^{2x}$$

2.92 Evaluate $\frac{df}{dx}$ at $x = 0$ for

$$f(x) = e^{(x^2)}$$

2.93 Evaluate $\frac{df}{dx}$ at $x = 0$ for

$$f(x) = \frac{1}{\left(x^2 - x\right)}$$

2.94 Evaluate $\frac{df}{dx}$ at $x = 1$ for

$$f(x) = \frac{-1}{x^2}$$

2.95 Evaluate $\frac{df}{dx}$ at $x = 0$ for

$$f(x) = \sin\left(x^2\right)$$

2.96 Evaluate $\frac{df}{dx}$ at $x = 1$ for

$$f(x) = \cos(2x)$$

2.97 Evaluate $\frac{df}{dx}$ at $x = 1$ for

$$f(x) = \ln(x)$$

2.98 Evaluate $\frac{df}{dx}$ at $x = 0$ for

$$f(x) = \sqrt{2x^3 + 7}$$

2.99 Evaluate $\frac{df}{dx}$ at $x = 0$ for

$$f(x) = (5x + 2)^3$$

2.100 Evaluate $\frac{df}{dx}$ at $x = 1$ for

$$f(x) = \left(x^3 + 1\right)\left(x^2 + 1\right)$$

Further Reading

1. Blitzer, R. (2017). *College algebra* (7th ed.). Pearson.
2. Briggs, W. L., Cochran, L., & Gillett, B. (2012). *Calculus for scientists and engineers, multivariable* (1st ed.). Pearson.

3. Briggs, W. L., Cochran, L. Gillet, B., & Schulz, E. (2012). *Calculus for scientists and engineers: Early transcendentals* (1st ed.). Pearson.
4. Brokate, M., Manchandra, P., & Siddiqi, A. H. (2019). *Calculus for scientists and engineers (Industrial and applied mathematics)* (1st ed.). Springer.
5. Kline, M. (1998). *Calculus: An intuitive and physical approach* (2nd ed.). Dover Publications.
6. Lial, M. L., Hornsby, J., Schneider, D. I., & Daniels, C. (2016). *College algebra* (12th ed.). Pearson.
7. Miklavcic, S. J. (2019). *An illustrative guide to multivariable and vector calculus* (1st ed.). Springer.
8. Miller, J., & Gerkin, D. (2016). *College algebra (Collegiate Mathematics)* (2nd ed.). McGraw-Hill Education.
9. Stewart, J., Redlin, L., & Watson, S. (2015). *College algebra* (7th ed.). Cengage Learning.

Chapter 3
Classification of Numerical Problems

3.1 General Classifications of Numerical Problems

Is the exact solution expected? Does the system change with time? How many unknowns are there? Is the system linear? Does the problem include rates of change? How much uncertainty is involved? These are general questions to consider when first approaching a problem.

3.1.1 Analytical vs. Numerical

Consider finding roots to an equation. In the simplest case, you have a specific function of one variable, $y = f(x)$, and you are searching for the values of x where $y = f(x)$ crosses or touches the horizontal line $y = 0$.

Most are more familiar with analytical (exact) solutions. For some problems, the solution has a closed-form exact solution. One example is the use of the quadratic formula to find the roots of a second-order polynomial with constant coefficients. In some cases, a solution may have an irrational numerical value, like $x = \sqrt{2}$. This solution may be approximated numerically as $x \approx 1.4142135$, but the solution is an exact value. In other cases, the analytical solution to a problem may be expressed using variable or parameters such as $x = \sqrt{c}$. The numerical solution for x given by the analytical expression $x = \sqrt{c}$ is not known until the variable c is assigned a value. There is a relationship between c and x, which can be understood in the general sense that increasing c will also increase the value of x.

> **Analytical Solution**—Analytical solutions are exact mathematical solutions to a given problem. Analytical solutions may be expressed in terms of numbers or unknown variables.

© Springer Nature Switzerland AG 2022
E. Gatzke, *Introduction to Modeling and Numerical Methods for Biomedical and Chemical Engineers*, https://doi.org/10.1007/978-3-030-76449-4_3

Numerical solutions are approximate solutions. A numerical solution may be a very good approximation to the exact solution; however, a numerical solution (almost) always includes some error. Understanding that numerical solutions include some error and how to deal with that error can be key to solving problems.

> **Numerical Solution**—Numerical solutions are approximate solutions to a mathematical problem, computed using actual numerical values for most of the values in the problem.

Once again, consider root finding. There are a variety of numerical methods to find roots of single equation. You may have a very complex expression, $f(x)$, that is not easily factored into roots or solved directly. To solve $f(x) = 0$, you can graph the expression and then examine the graph to locate *zero crossings* at the values of x that satisfy the function. Using the bisection method, you can evaluate the function at two points, x_L and x_R, $x_R \geq x_L$. Assuming that $f(x_L) \leq 0$ and $f(x_U) \geq 0$, you know a root must lie in the region $x_L \leq x \leq x_R$. Bisect the region to find $x_M = x_L + \frac{x_R - x_L}{2}$ and evaluate the function at x_M. You can update the bounds on x, keeping the region that must contain a solution. Using either a graphical method or iterative method will result in an answer that is not exact.

It is important for engineers to understand that both analytical and numeric answers are important and useful in different ways.

3.1.2 Steady-State vs. Dynamic

Real systems change with time. In many cases, engineers may assume that the rate of change is not significant in solving the problem. When something is not changing very rapidly, it is assumed to be at "steady-state." In some cases, problems are assumed to be steady-state because solution of the dynamic problem may be too complicated.

> **Steady-State**—A system is at steady-state if it does not change with time.

Many problems do change with time. Materials heat up, reactions take place, cells grow, or other changes may make the rate of change significant.

> **Dynamic**—A system is dynamic if it changes with time.

Dynamic systems usually require more advanced mathematics. If at all possible, engineers will often start with simpler steady-state models to help determine a basic solution. If warranted, a dynamic model and solution could be developed.

In chemical and biomedical engineering, many models are based on a material balance on a volume. If positive and negative rates exactly balance, the rate of accumulation for the volume under consideration is zero. The process is assumed to be at steady-state, so dynamic solution may not be needed.

3.1.3 Scalar vs. Multivariable Algebraic Equations

Engineers use equations to specify mathematical relationships between various variables. The simplest case one may work on involves a single variable. This is called a scalar value.

> **Scalar**—A scalar is a single number. A scalar equation only involves a single variable.

A mathematical function provides a mapping. A *scalar* function maps one value to another value. A function like $y = f(x) = x^3$ can take any real value for x and produce any real value for y. This function could be described as $f(x) : \mathbb{R}^1 \to \mathbb{R}^1$. The notation $\mathbb{R}^1$ means all scalar real numbers and the 1 denotes dimension 1. A function can be considered an input–output relationship. The "black-box" analogy is commonly used; you input some number (the *independent variable*) and another comes out (the *dependent variable*). A function is an input–output relationship (and should pass the vertical line test), while an equation relates the two expressions on the left and right side of the equality sign.

Every engineer should understand the basic concept of a mathematical function or an algebraic expression. Consider a single pressure tank that is leaking or filling with air. The tank pressure is changing with time, so the pressure in a tank, P, could be some function of time, $P(t)$ or $P(t) = f(t)$, or in a specific case, $P(t) = 5e^{-3t} + 14.7$. The pressure variable is a scalar value, despite it being a function of time. Some examples that are functions of time are:

$$f(t) = \sin(5\,t)$$
$$f(t) = 2\,t$$
$$f(t) = e^{3t}$$

In other cases, one might consider a variable that changes with temperature. Consider a chemical reaction rate. This rate could be expressed as $r(T) = k_o e^{-\frac{E}{RT}}$, where E, k_o, and R are assumed constant. This means that the only variable value is temperature T. This is still a scalar function despite including multiple apparent variables. If those values are held constant, there is only one degree of freedom for the expression: changing the temperature T.

3.1.3.1   Multivariable Algebraic Equations

A value could be a function of multiple values. Again, the mathematical function provides a mapping. A function can map multiple values to another value. $f(x)$: $\mathbb{R}^n \rightarrow \mathbb{R}^1$, $n > 0$. The notation $\mathbb{R}^n$ means the space defined by n scalar real numbers. Usually, we think about 2D and 3D spaces, but some mathematics requires we deal with functions in higher dimensional spaces that we cannot visualize as easily.

Multivariable Equation—A multivariable equation involves multiple variables.

Consider the ideal gas law for a constant amount of gas in a cylinder:

$$PV = nRT$$

This can be rearranged to solve for pressure such that:

$$P = \frac{nRT}{V}$$

Assuming that the gas constant is constant, pressure P depends on three values: the number of moles n, the volume V, and the temperature T. Pressure could be written as a function of three variables:

$$P(n, T, V) = \frac{nRT}{V}$$

Multiple equations would be required to actually solve for the unknown solution. Given n variables, you generally need n independent algebraic equations in order to find a solution. There are sets of equations that have no solution and some that have multiple solutions.

Example 3.1 (Multivariable Functions) Consider a topographic (elevation) map where the elevation z of the land is a function of map position given by the x and y coordinates:

$$z = f(x, y) = z = -(x^2 + y^2) + 100$$

To determine the distance on a 2D plane from a point, one must use the distance formula. Consider the distance from the point (4, 2):

$$d = f(x, y) = \sqrt{(x - 4)^2 + (y - 2)^2}$$

Another example could be a complex reaction rate expression that is a function of reactant concentrations C_A and C_B as well as temperature T:

$$r = f(C_A, C_B, T) = 3.0\,e^{-\frac{3}{8.14T}}\,C_A^2\,C_B$$

Multivariable System of Algebraic Equations—A multivariable system involves multiple algebraic equations and variables.

Example 3.2 (Multivariable System) Consider two multivariable equations. First, define a circle of radius 3:

$$x^2 + y^2 = 9$$

Next, consider a simple parabola:

$$y = x^2$$

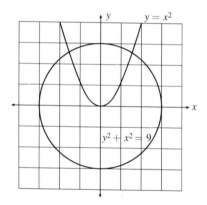

This can be written as a system of equations:

$$f_1(x, y) = x^2 + y^2 - 9$$
$$f_2(x, y) = y - x^2$$

In this case, there are two points where the functions intersect. The two equations are shown in the figure above. To satisfy both equations, you must find the point (x, y) that satisfies both equations, making $f_1(x, y) = f_2(x, y) = 0$ at that point.

3.1.4   Linear vs. Nonlinear

Two major classes of equations are **linear** and **nonlinear** equations.

Linear Algebraic Equation—A linear algebraic equation can be expressed for N variables x_i and $N + 1$ constants c_i as:

$$\sum_{i=1}^{N} c_i x_i = c_0$$

In many cases, analytical (exact) and numerical (approximate) methods work only for linear equations. In other cases, a linear approximation may be used to find approximate solutions using linear methods. Linear equations are generally easier to solve. Engineers will often work to formulate a problem with linear equations or make assumption to make the problem linear. This leads to the following less useful definition for nonlinear equations.

Nonlinear Algebraic Equation—A nonlinear equation is any equation that is not linear.

Some analytical and numerical methods can be used with nonlinear equations. Generally, nonlinear equations require more complex solution methods. Some simple examples of a nonlinear equation include a simple parabola $y = a + bx^2$ and the exponential function $y = e^{cx}$.

3.1.5   Algebraic vs. Differential

Most of your mathematical education has focused on simple relationships relating variables. These algebraic relationships can often be solved using basic algebra.

Algebraic Equation—A simple equality relationship between variables.

Some examples of algebraic relationships include:

- $PV = nRT$ the ideal gas law,
- $\rho = \frac{m}{V}$ the density relationship, and
- $V = iR$ Ohm's law.

With the development of differential calculus, many new problems could be formulated and solved. These equations often include derivatives representing the rate of change of a variable as an unknown value.

> **Differential Equation**—A relationship between variables involving the derivative of one or more variables.

Solving an algebraic problem for unknown variable values will typically lead to a single value for each variable. Solving a differential equation for an unknown variable will give an expression for that variable as a function of some independent variable such as space or time. Algebraic solutions may yield a value for pressure P or concentration C. Solutions to a differential equation will give an expression for pressure as a function of time $P(t)$ or concentration as a function of position $C(x)$.

Example 3.3 (Ballistics) Consider the basic physics equations relating force F, mass m, and acceleration a:

$$F = ma$$

This appears to an algebraic equation and can often be treated as such. To determine the weight of a rock on the moon, one would need the mass of the rock and the acceleration due to the gravity of the moon.

However, acceleration is defined as the rate of change of velocity v, leading to:

$$F = ma = m\frac{dv}{dt}$$

Velocity is defined as the rate of change of position. Consider a one-dimensional problem, where x represents the object position:

$$v = \frac{dx}{dt}$$

resulting in two differential equations to be solved to determine the position and velocity as a function of time:

$$a = \frac{dv}{dt}$$

$$v = \frac{dx}{dt}$$

Given initial conditions and a value for the acceleration a, these differential equations could be solved to find algebraic expressions for $v(t)$ and $x(t)$.

3.1.6 Deterministic vs. Stochastic

Engineers must deal with real systems in the real world. To model these systems, engineers write equations that represent the system. These models are assumed to be accurate. However, when compared to real data, the data may appear noisy, while the mathematical model does not randomly change.

> **Deterministic**—A deterministic value is assumed to not randomly change.

Noise and variation comes from many sources. Electronics can be affected by background fluctuations. Biomedical problems are affected by population randomness. Chemical reactions at the molecular level can be represented with a probability of occurrence.

> **Stochastic**—A stochastic value randomly changes.

Engineers must learn how to handle stochastic systems despite only having deterministic models. For treatment of data, computing the average and variance helps quantify the measurement uncertainty. In some cases, a model can be used in a random simulation run multiple times to help quantify expected uncertainty.

3.2 Common Classes of Numerical Problems

Engineering problem solving is really about pulling together concepts from science and methods from mathematics to formulate a problem and arrive at a solution. Once a problem is formulated, it is up to the engineer to determine what type of problem it is and how it could be solved. This chapter provides a variety of example problems that fall into general classes.

3.2.1 Algebraic Equations

Most concepts are described by simple algebraic relationships. For example, the density of an object is defined as the ratio of its mass to its volume:

$$\rho = \frac{m}{V}$$

In physics, particle motion is often simplified to the equation:

$$F = m\,a$$

Many gasses follow the ideal gas law in some situations. This relates the pressure of the gas, the volume of the gas, the number of moles of the gas, and the temperature of the gas.

$$PV = nRT$$

These algebraic equations include unknown variables. The equations may also include constant values, like R, the universal gas constant.

Basic rules of algebra can be used to manipulate algebraic equations. Some equations can easily be rearranged to make a variable explicitly defined. For the ideal gas law, solving for P defines the pressure explicitly in terms of the number of moles of the gas, the temperature, and the volume:

$$P = \frac{nRT}{V}$$

Some algebraic equations involving a single unknown are difficult or even impossible to solve using basic algebra. Instead of solving the equation for an exact solution, numerical methods such as Newton's method (Chap. 8) can be used to determine approximate numerical solutions for an unknown value.

In some cases, problems may involve multiple algebraic equations involving linear relationships. These types of problems lend themselves to matrix methods discussed in Chap. 7. If all the equations are linear, matrix methods can be used to solve problems involving thousands of unknown variables.

In more complicated cases, multiple nonlinear algebraic equations can be solved using a multivariable version of Newton's method. The multivariable Newton–Raphson method discussed in Chap. 8 uses an iterative approach, approximating the set of nonlinear equations as linear equations and then solving the set of linear equations at each step.

Typically, you must have N algebraic relationships to solve for N variables to find a unique solution. There are cases where you may have infinitely many solutions:

$$y = x$$
$$2y = 2x$$

where the two equations obviously overlap at every point. There are cases where you may have no solution:

$$y = x$$
$$y = x + 1$$

and no variable values can be found to satisfy both relationships. When discussing algebraic relationships, the number of variables in a problem is often referred to as the degree of freedom for the problem.

Degrees of Freedom—The number of independent variables in a system of equations.

If a problem has zero degrees of freedom, there could be a solution. In some cases with zero degrees of freedom, there are no solutions. For example, the following set of two nonlinear equations with two unknown variables has no real-valued solution:

$$x^2 + y^2 = 9$$
$$y = x^2 + 4$$

If a problem has one degree of freedom, one additional relationship must be incorporated so that the problem is well-posed. For linear algebraic systems of equations, one may calculate the determinant to determine if a system of equations is well-posed. There is no readily available general-purpose method for checking systems of nonlinear algebraic equations to see if a solution is possible.

3.2.2 Numerical Optimization

A special class of problem arises when inequalities are part of the problem or there are available degrees of freedom. Limits on variables or one-side relationships lead to an under-specified problem with degrees of freedom. These limits often arise from real-world considerations like "concentrations must be non-negative" or "the reactor temperature must remain below the design limit value."

Consider the space defined by the following four equations:

$$x \geq 1$$
$$y \geq 1$$
$$y \leq -\frac{1}{2}x + 1$$
$$y \leq -2x + 2$$

These inequalities define a region in the $x-y$ plane. Any point inside the feasible region could be a solution. The feasible space as the intersection of the inequality constraints is shown in Fig. 3.1.

In some cases, constraints are developed where no points can be found to satisfy all of the constraints. These problems arise from contradictory constraints.

Fig. 3.1 Feasible space for
four inequality constraints.
Given a desire to maximize
$x + y$, the improving
direction is shown at the point
$x = 2, y = 2$

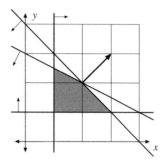

In engineering design problems, this may require that some constraints be modified
or "loosened." Problems without a solution are described as infeasible.

Infeasible—A multivariate optimization problem where no point may be found
to satisfy all of the specified constraints.

In addition to the relationships on the feasible variable space, a problem may also
include an objective function to be maximized or minimized. Perhaps it is desired to
maximize the sum of the two variables, such as the quantity $x + y$. This is equivalent
to minimization of $-x - y$. The goal of the optimization problem is to find a feasible
point that satisfies all constraint relationship while minimizing (or maximizing) the
specified objective function. Using concepts from vector calculus, the improving
direction can be determined. The improving direction to maximize $x + y$ is shown
in Fig. 3.1 at the optimal solution point: $x = 2, y = 2$.

3.2.3   Ordinary Differential Equations

In some cases, engineering relationships are derived using rates. Instead of having
a relationship with just variables, the relationship includes the rate of change of one
or more variables. This inclusion of rate of change results in differential equations.
Solving a differential equation for variable of interest usually gives a solution that
depends other variables.

Consider the tank draining problem for a tank of constant cross-sectional area A
and flow out proportional to the height, $kh(t)$. These assumptions result in a simple
linear differential equation:

$$A\frac{dh}{dt}(t) = -k\,h(t)$$

In this problem, the tank height $h(t)$ is unknown. The ***initial condition*** for the
height is assumed to be h_0 at time $t = 0$. The solution to this specific differential

equation with known initial conditions can be derived as:

$$h(t) = e^{-\frac{k}{A}t} h_0$$

This means that the value of the tank height is a function of time. This algebraic equation solution "satisfies" the differential equation for all time values ≥ 0. You can check by taking the derivative of the solution function $h(t)$ with respect to time:

$$\frac{d}{dt}(h(t)) = -\frac{k}{A}\left(e^{-\frac{k}{A}t}\right) h_0$$

Substituting this value for $\frac{dh}{dt}(t)$ and the expression for $h(t)$ back into the original differential equation:

$$A\frac{dh}{dt}(t) = -k\,h(t)$$

$$A\left(-\frac{k}{A}\left(e^{-\frac{k}{A}t}\right) h_0\right) = -k\left(e^{-\frac{k}{A}t} h_0\right)$$

shows that the proposed function is a valid solution.

> **Ordinary Differential Equations**—Ordinary Differential Equations (ODEs) only depend on one variable.

The solution to an ODE will be parameterized as a function of a separate variable. This is generally called the independent variable. In the previous example, time t is the independent variable. Other independent variables are also used in engineering. The velocity of fluid flow in a tube or pipe at steady-state is a function of axial position r. In a simple tubular reactor model, the concentration depends on the position along the length of the tube, $C(x)$. The key element for an Ordinary Differential Equation is that the solution only depends on one independent variable.

In an ODE problem, there may be many different equations. Consider a tank draining into another tank. There will be a differential equation for each tank. One tank depends on the other. The solution for the tank heights will still only depend on one independent variable, time. Consider two identical tanks draining with identical valves:

$$A\frac{dh_1}{dt}(t) = -k\,h_1(t)$$

$$A\frac{dh_2}{dt}(t) = k\,h_1(t) - kh_2(t)$$

The solution for $h_1(t)$ is the same as before. However, the solution for $h_2(t)$ is more complicated because it depends on $h_1(t)$. The solution is:

$$h_1(t) = e^{-\frac{k}{A}t}\, h_1(0)$$

$$h_2(t) = e^{-\frac{k}{A}t}\left(k\,t\,h_1(0) + \frac{h_2(0)}{A}\right)\frac{1}{A}$$

Both $h_1(t)$ and $h_2(t)$ only depend on time. Initial conditions for the tanks and values for parameters k and A can be used to find the predicted tank heights at any point in time.

Many ODEs are initial value problems where only the variable initial values must be known to find the solution. Sometimes, initial value problems are called shooting problems, taking a term from ballistics where the initial velocity is known and the trajectory is determined. Some ODEs are boundary value problems where the variable must satisfy certain limits at one boundary. For the fluid in a pipe, the fluid velocity is assumed to be 0 at the wall and the fluid velocity rate of change is assumed to be 0 at the center of the pipe.

$$v(R) = 0, \quad \frac{dv}{dx}(0) = 0$$

ODEs are often described by their order. A second-order ODE could involve two related equations, each with a first derivative. Alternatively, a second-order ODE may be a single equation involving a second derivative expression.

3.2.4 Partial Differential Equations

Partial Differential Equations (PDEs) are more complicated than ODEs.

> **Partial Differential Equations**—Partial Differential Equations (ODEs) depend on more than one independent variable.

Consider the temperature profile in a solid piece of metal. The temperature at every point may be different from every other point. A steady-state solution will depend on x, y, and z. The solution variable for the temperature $T(x, y, z)$ could also depend on time, $T(x, y, z, t)$ if the object is not assumed to be at steady-state.

Simplifying assumptions are often made. For example, the piece of metal could be assumed to be a single brick in a furnace exposed to heat on one side and cold on the other side. Assuming that there are enough bricks surrounding the brick of interest, the dependence on y and z could be removed. Now, the solution for temperature would only depend on x and t, $T(x, t)$.

Consider fluid flowing in a pipe or tube. At steady-state, the velocity of the fluid only depends on axial position r, $v(r)$. A pipe flowing out of a large tank has a more complicated flow profile that also depends on distance from the tank wall, x.

The flow profile solution would then become $v(x, r)$. Time dependence could be included as well, making the solution $v(x, r, t)$.

A major consideration for PDEs is the specification of initial conditions and boundary conditions. For fluids, "no-slip" boundary conditions mean that the fluid has a zero velocity at the wall. For heat problems, the temperature value at a boundary could be specified. Alternatively, the rate of heat passing through the boundary (heat flux) could be specified. For a well-insulated wall, the flux is 0. For a solid–liquid or solid–gas interface, there may be a heat flux expression such as Newton's law of cooling.

3.2.5   Differential Algebraic Equations

Differential Algebraic Equations (DAEs) are similar to ODEs in many ways. In many engineering problems, Ordinary Differential Equations can be written where all the dependent variables are functions of time, the only dependent variable. These equations may derive from balances with accumulation rates. However, additional relationships may require inclusion of algebraic equations that only include variables, not the time derivative of variables. In some cases, algebra cannot be used to solve explicitly for the derivative variables. Special simulation and solution methods must be used.

Consider three pressure tanks connected as shown in Fig. 3.2, Case A. Molar balances can be written for each tank. Using the ideal gas law $PV = nRT$, the molar accumulation rate $\frac{dn}{dt}$ can be expressed as $\frac{V}{RT} \frac{dP}{dt}$ if the tank volume and gas temperatures are assumed constant. Taking molar flow rates across the valves as $F_{ij}(t)$, three Ordinary Differential Equations could be written for this system as:

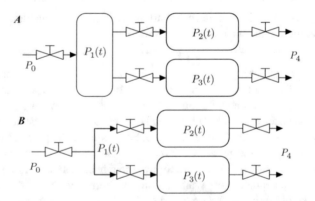

Fig. 3.2 Two different tank configurations. Case A may be modeled with three ODEs, while case B includes an algebraic equation for determination of pressure $P_1(t)$

$$\frac{V_1}{RT}\frac{dP_1}{dt}(t) = F_{01}(t) - F_{12}(t) - F_{13}(t)$$

$$\frac{V_2}{RT}\frac{dP_2}{dt}(t) = F_{12}(t) - F_{24}(t)$$

$$\frac{V_3}{RT}\frac{dP_3}{dt}(t) = F_{13}(t) - F_{34}(t)$$

The upstream pressure P_0 is constant, as is the atmospheric pressure P_4. Assuming that the molar flow rate across a valve is proportional to the square root of the pressure difference: $F_{ij}(t) = k_{ij}\left(P_i(t) - P_j(t)\right)^{1/2}$, the three equations can be written as:

$$\frac{V_1}{RT}\frac{dP_1}{dt}(t) = k_{01}\left(P_0 - P_1(t)\right)^{1/2} - k_{12}\left(P_1(t) - P_2(t)\right)^{1/2} - k_{13}\left(P_1(t) - P_3(t)\right)^{1/2}$$

$$\frac{V_2}{RT}\frac{dP_2}{dt}(t) = k_{12}\left(P_1(t) - P_2(t)\right)^{1/2} - k_{24}\left(P_2(t) - P_4\right)^{1/2}$$

$$\frac{V_3}{RT}\frac{dP_3}{dt}(t) = k_{13}\left(P_1(t) - P_3(t)\right)^{1/2} - k_{34}\left(P_3(t) - P_4\right)^{1/2}$$

The three independent variables are the tank pressures, $P_1(t)$, $P_2(t)$, and $P_3(t)$. This represents three nonlinear coupled Ordinary Differential Equations. The initial conditions at time $t = 0$ are $P_1(t = 0)$, $P_2(t = 0)$, and $P_3(t = 0)$.

Now, consider the case where tank 1 is quite small, $V_1 \ll V_2$ and $V_1 \ll V_3$. This is shown in Fig. 3.2, Case B. This means that $P_1(t)$ approaches steady-state much more rapidly than $P_2(t)$ and $P_3(t)$. Letting the rate of change for $P_1(t)$ approach 0 as V_1 approaches 0 leads to:

$$0 = k_{01}\left(P_0 - P_1(t)\right)^{1/2} - k_{12}\left(P_1(t) - P_2(t)\right)^{1/2} - k_{13}\left(P_1(t) - P_3(t)\right)^{1/2}$$

$$\frac{V_2}{RT}\frac{dP_2}{dt}(t) = k_{12}\left(P_1(t) - P_2(t)\right)^{1/2} - k_{24}\left(P_2(t) - P_4\right)^{1/2}$$

$$\frac{V_3}{RT}\frac{dP_3}{dt}(t) = k_{13}\left(P_1(t) - P_3(t)\right)^{1/2} - k_{34}\left(P_3(t) - P_4\right)^{1/2}$$

Here, the first equation is an algebraic equation that must be satisfied at all time values. This equation cannot easily be solved for $P_1(t)$ as a function of $P_2(t)$ and $P_3(t)$. Specialized numerical solution methods may be needed. Just finding initial conditions that satisfy the algebraic functions for DAE systems can be quite challenging.

If the flow rate expression were a simple linear relationship ($F = k\Delta P$), the linear equation could be solved for $P_1(t)$ as a function of $P_2(t)$ and $P_3(t)$ as follows:

$$0 = k_{01}\left(P_0 - P_1(t)\right) - k_{12}\left(P_1(t) - P_2(t)\right) - k_{13}\left(P_1(t) - P_3(t)\right)$$

$$P_1(t)\left(-k_{01} + k_{12} + k_{13}\right) = k_{01} P_0 + k_{12} P_2(t) + k_{13} P_3(t)$$

$$P_1(t) = \frac{k_{01} P_0 + k_{12} P_2(t) + k_{13} P_3(t)}{\left(-k_{01} + k_{12} + k_{13}\right)}$$

This expression can allow the two differential equations to be expressed only in terms of P_2 and P_3, making this an ODE system with only two variables: $P_2(t)$ and $P_3(t)$.

$$\frac{V_2}{RT}\frac{dP_2}{dt}(t) = k_{12}\left(\left(\frac{k_{01} P_0 + k_{12} P_2(t) + k_{13} P_3(t)}{\left(-k_{01} + k_{12} + k_{13}\right)}\right) - P_2(t)\right) - k_{24}\left(P_2(t) - P_4\right)$$

$$\frac{V_3}{RT}\frac{dP_3}{dt}(t) = k_{13}\left(\frac{k_{01} P_0 + k_{12} P_2(t) + k_{13} P_3(t)}{\left(-k_{01} + k_{12} + k_{13}\right)} - P_3(t)\right) - k_{34}\left(P_3(t) - P_4\right)$$

Problems

For the following problems, do NOT attempt to solve for an actual solution.

1. For the following problem, identify the unknown variable or variables. How many equations are there? Are the equations algebraic, differential, or differential/algebraic? Are the equations linear or nonlinear?

$$2x + 1 = 4$$

2. For the following problem, identify the unknown variable or variables. How many equations are there? Are the equations algebraic, differential, or differential/algebraic? Are the equations linear or nonlinear?

$$3x^2 + 1x + 2 = 6$$

3. For the following problem, identify the unknown variable or variables. How many equations are there? Are the equations algebraic, differential, or differential/algebraic? Are the equations linear or nonlinear?

$$2x + 4y = 9$$
$$y - x = 8$$

4. For the following problem, identify the unknown variable or variables. How many equations are there? Are the equations algebraic, differential, or differential/algebraic? Are the equations linear or nonlinear?

$$2x^{2.3} - 7 = 4$$

5. For the following problem, identify the unknown variable or variables. How many equations are there? Are the equations algebraic, differential, or differential/algebraic? Are the equations linear or nonlinear?

$$y = 2x^2$$
$$x^2 + y^2 = 4$$

6. For the following problem, identify the unknown variable or variables. How many equations are there? Are the equations algebraic, differential, or differential/algebraic? Are the equations linear or nonlinear?

$$\frac{dx}{dt}(t) = -3x(t)$$

7. For the following problem, identify the unknown variable or variables. How many equations are there? Are the equations algebraic, differential, or differential/algebraic? Are the equations linear or nonlinear?

$$4\frac{dx}{dt}(t) = x(t) - (x(t))^2$$

8. For the following problem, identify the unknown variable or variables. How many equations are there? Are the equations algebraic, differential, or differential/algebraic? Are the equations linear or nonlinear?

$$\frac{dx_1}{dt}(t) = -3x_1(t) - 2x_2(t)$$

$$\frac{dx_2}{dt}(t) = -x_2(t)$$

9. For the following problem, identify the unknown variable or variables. How many equations are there? Are the equations algebraic, differential, or differential/algebraic? Are the equations linear or nonlinear?

$$\frac{dx_1}{dt}(t) = -x_1(t) + x_3(t)$$

$$\frac{dx_2}{dt}(t) = -x_3(t)$$

$$0 = (x_1(t) - x_3(t))^2 + \frac{x_1(t)}{x_3(t)}$$

10. For the following problem, identify the unknown variable or variables. How many equations are there? Are the equations algebraic, differential, or differential/algebraic? Are the equations linear or nonlinear?

$$\frac{\partial T}{\partial x}(x) = -3T(x)$$

Chapter 4
Engineering Modeling

4.1 Forces and Free Body Diagrams

Most people are familiar with basic mechanical systems. Everyone experiences gravity, either falling down or throwing a ball. This gravitational force acts on all objects on Earth. For analysis of mechanical systems, forces act on the system to change the motion of an object.

> **Force**—A force pushes or presses on a mass to potentially change the inertia of the object.

A force has both direction and magnitude. This will be discussed more in Chap. 13. Forces can be due to gravitational attraction, magnetic attraction/repulsion, or even mechanical contact with something pressing on an object. For simple systems, the gravitational force is assumed to act on the center of gravity.

> **Center of Gravity**—The center of mass of an object.

The center of gravity simplifies analysis and modeling. To properly determine interactions between an object and related forces could require complex mathematical analysis, integrating the effect of forces over the entire object. For initial analysis, engineers often use a free body diagram.

> **Free Body Diagram**—A schematic which includes all relevant forces acting on an object.

The free body diagram shows what forces are acting on an object. The free body diagram also may indicate the dimensions of the object and the location

© Springer Nature Switzerland AG 2022

E. Gatzke, *Introduction to Modeling and Numerical Methods for Biomedical and Chemical Engineers*, https://doi.org/10.1007/978-3-030-76449-4_4

and direction of any applied forces. In simple cases, the forces on an object balance at equilibrium, causing the object to either not move or change its velocity. Mathematically, this is expressed as:

$$\Sigma F_i = 0$$

In more advanced situations, forces may induce a rotational motion or moment in the object. Forces and moments will be considered in more detail in Chap. 13.

4.1.1 Simple Force Balance Examples

Example: Falling Droplet at Terminal Velocity

Consider a droplet of liquid falling freely. This droplet is subject to gravity, which provides a downward force. At terminal velocity, the droplet will fall at a constant rate toward the ground. This means it is not accelerating. This means that the force of gravity must be balanced. In this case, a drag force in the upward direction acts on the droplet. This force due to wind resistance will balance the gravitational force. Also, wind resistance can be modeled as a function of the droplet velocity and shape, allowing for analytical solution of the terminal velocity under some conditions.

Example: Buoyant Floating Solid

Consider a hollow object at rest floating in a liquid. The total weight of the object provides the downward gravitational force. If the object is at rest, there must be a force balancing the downward force of gravity. In this case, the liquid pushes up against the object. This force is called buoyancy. The density of the object can be determined given the weight and volume of the object. The object will only float if the effective density of the hollow object is less than that of the liquid. Given the density of the liquid, the total volume of liquid displaced can be determined by realizing the weight of the displace liquid will balance the weight of the hollow object.

4.2 Electric Circuits

Electricity has enabled the modern age. Electricity allows for so many different machines to do mechanical work. Using electricity, energy may be transferred from one place to another, avoiding mechanical, hydraulic, or pneumatic connections. Electricity also scales readily, allowing for creation of powerful machines. Modern technology relies on use of electricity in so many different ways.

To consider modeling electric systems, one must first understand something about current and voltage.

Current—Current is the movement of electrons.

Voltage—Voltage is the electric potential.

Everyone is familiar with batteries. Batteries transform stored chemical energy into electric energy. One single AA battery supplies electricity at 1.5 V. Putting four AA batteries in series should create a 6-volt potential. However, AA batteries are physically smaller than larger C and D cell batteries. AAA, AA, C, and D all generate 1.5 V but the larger C and D cells can supply more current. The electric power supplied depends on both voltage and potential.

Electric Power—Electric power is the rate at which electric power is converted to another form of energy.

Assuming voltage is expressed as variable V in volts and the current is expressed as variable I in amps, the electric power P is calculated as:

$$P = V I$$

Power is expressed in units of watts (W) or equivalently $\frac{\text{joule}}{\text{s}}$ or $\frac{\text{kg m}}{\text{s}^3}$.

Electric power may be compared with an analogy of water in a hose. The pressure of water in the hose is like the voltage or electric potential. The amount of water flowing in a hose is like the current: a garden hose moves less water than a fire hose. A water pump is like a battery in that it increases the pressure (voltage) in the hose and has both an inlet and outlet. A circuit may be formed when the water is pumped out to a turbine (electric motor or load) to do work then returns to the pump to be pressurized again. Valves can be seen as resistors in the circuit. This analogy is not perfect, but it often helps to visualize electrical systems with more physical real-world systems.

Most introductory electrical analysis considers only DC (Direct Current) applications where current is moving in one direction. However, we experience AC (Alternating Current) from our residential use of electricity. AC current moves electrons back and forth. Both AC and DC circuits can perform work, but AC has some advantages for safety and long range transmission efficiency. Analysis of AC circuits requires more advanced mathematics and is typically introduced in electrical engineering courses.

Gustav Kirchhoff formalized (at least) two significant relationships used to analyze electric circuits. One law describes a relationship between currents in a circuit and one law describes voltages in a circuit.

4.2.1   Kirchhoff's Current Law

Currents flowing into a node must equal the sum of the currents leaving a node.

$$\sum I_{in} = \sum I_{out}$$

Basically, the currents flowing into and out of an electric connection must balance. In circuit diagrams, a node may appear as a point connected to three or more wires. In a real system, wires make connections between components like resistors and LEDs. A node is the conductive material between components. If two or more components are electrically connected, the current entering and leaving the connection should balance.

4.2.2   Kirchhoff's Voltage Law

The sum of voltages in series must equal the overall voltage in a loop.

$$0 = \sum V_i$$

The voltages in an electric system must balance. Consider a battery which supplies 3 V of electric potential to single component, a resistor. The resistor is the only component in the loop, so the voltage drop across the resistor must be -3 V. This law is very useful for multiple components in series. See the example described in Sect. 1.2.2.

4.3   Conservation of Mass

This section considers modeling using conservation of mass as the fundamental concept. This means that under normal circumstances, matter is neither created or destroyed. Chemical reactions can take place where initial reactant species react to form product species by rearranging bonds to form new product molecules. These types of models are often referred to as "Mass Balances" although one should use expressions based on molar quantities. Molar concentrations may be calculated based on molar flow rates, molar reaction rates, and molar diffusion rates.

To consider chemically reactive systems, one must first understand how to quantify the molecules involved in a chemical reaction. The concentration of a species provides some information.

> **Concentration**—The concentration of a substance in a mixture represents how much of the substance is contained within a specific volume of the mixture. It is usually expressed as (moles of substance per volume of mixture).

The concentration value describes how much of a specific molecule is contained in a volume of the overall mixture or solution. This assumes that the substance under consideration is well-mixed. If the material being considered is well-mixed, it does not matter where a sample is taken from or how much sample is extracted; any sample will have the same concentration values when extracted from a well-mixed source.

The concentration of a substance is usually expressed as the number of moles of a material found in a given volume of the mixture. For example, a 2.0 molar HCl solution will have 2 moles of elemental hydrogen and 2 moles of elemental chlorine in each liter of the substance. This can be expressed as $2\ \frac{mol}{L}$.

In some cases, a concentration based on mass is used. Since a given number of moles of a substance will contain a set number of molecules, the conversion between moles and mass is easy if you know the molecular weight.

Example 4.1 (Concentration) A given solution of HCl is $1.3\ \frac{mol}{L}$. This means that in one liter of solution, there are 1.3 moles of the molecule under consideration. For each HCl molecule, there will be one atom of hydrogen and one atom of chlorine. Given that the atomic weight of hydrogen is 1.01 and the atomic weight of chlorine is 35.5, the mass based volume can be calculated as:

$$1.3\ \frac{mol\ HCl}{L} \times \frac{1.01\,g\,H + 35.5\,g\,Cl}{1\ mol\ HCl} = 47.5\ \frac{g\,HCl}{L}$$

This means that in every liter of the solution there are 47.5 g of HCl.

In many cases, concentrations of a substance are considered for a liquid form. In water-based mixtures, this is also called the aqueous concentration. However, one may also consider gaseous mixtures. A well-mixed volume of a mixture of gasses will also have a concentration for the various components in the mixture.

To consider modeling of various reactive systems, the concept of a control volume must be introduced.

> **Control Volume**—A control volume defines a boundary when modeling a chemical system. The control volume is sometimes assumed to be well-mixed, meaning that the concentration of a substance in the control volume is the same throughout the entire volume.

In a simple chemical reactor, the control volume would match the volume of the reactor. However, there are many types of complicated reactors. Consider a reactor filled with small porous catalyst pellets where the reaction takes place inside the small pellets. To accurately model this reactor, it may be necessary to model the reactor volume outside the catalyst pellets as one control volume while separately considering the volume inside the pellets as a separate control volume.

Consider the most complicated reactor found in nature, the cell. Inside the cell wall, a reaction could take place inside the cytoplasm. The cell contents may not be well-mixed, but for engineering purposes it may be an adequate representation of the overall population of cells. Alternatively, more complicated cell models can be developed, considering other control volumes such as the cell wall concentrations, organelle concentrations, or nuclear concentrations.

Given the concepts of molecular concentration and control volume, it makes sense to proceed to describing the chemical changes that are taking place.

> **Chemical Reaction**—A chemical reaction transforms one or more reactant molecules into one or more product molecules.

As stated previously, chemical reactions break and form chemical bonds between the atoms in a molecule to form new and different molecules. Unless a nuclear reaction takes place, matter is not created or destroyed in a chemical reaction.

Reactions typically take place under specific conditions. The pressure and temperature may need to have certain values for the chemical reaction to rapidly take place. In some cases, a catalyst must be present to help the reaction. A catalyst participates in a reaction but is not consumed in the reaction. In biological systems, enzymes catalyze biochemical reactions.

In a chemical reaction, there are sometimes many reactants and many products. We know that matter cannot be created or destroyed. This means that the total number of atoms of one species before the reaction occurs must balance the total number of atoms of that same species after the reaction proceeds. This concept is used to balance chemical reactions.

> **Stoichiometry**—The stoichiometry of a chemical reaction defines how many of each reactant molecule is consumed to produce various numbers of product molecules.

The number of reactant or product molecules in a balanced chemical reaction are termed stoichiometric coefficients. The stoichiometry of a reaction helps determine what is going on inside the reaction. If two molecules of species A are required to react together to make molecule B, the balanced reaction would be:

$$2A \rightarrow B$$

This means that every time this reaction occurs, two molecules of A combine to make a single molecule of B. Knowing how often this reaction occurs would tell you how often species B is created. However, species A is consumed twice as fast because of the stoichiometry of the reaction. The entire reaction depends on the reaction rate.

Reaction Rate—The reaction rate is the rate that a chemical reaction proceeds. This value is usually expressed as (moles of a reacting molecule per time).

While chemical reactions often proceed very rapidly, they are not instantaneous. Understanding the rate of reaction can be a key engineering value to consider. If a reaction produces a high-value polymer, it would be desirable to increase the reaction rate so that as much polymer can be produced in a given reactor. However, allowing a chemical reaction to proceed too rapidly can result in a catastrophic explosion. In biological systems, the reaction rate could describe the rate of fermentation in a bioreactor where sugars are converted by yeast to alcohol. Alternatively, in a patient with Alzheimer's, the rate of plaque formation in neuron cells would be undesirable. It is therefore necessary to understand reaction rates.

Volumetric Reaction Rate—The volumetric reaction rate is the reaction rate per unit volume at which a chemical reaction proceeds. This value is usually expressed as (reaction rate per volume).

Rather than just quantify the total amount of a species that is reacted, one should consider the volumetric reaction rate. Given a volumetric reaction rate allows for easy use in many modeling scenarios. If the volumetric reaction rate expression is known, it can be used in any size reactor. The same reaction rate expression could hold true for a table-top micro-reactor just as readily as 50,000 gallon industrial reactor.

Chemical reactions often take place in a batch reactor. In a batch reactor, reactants are added, the reactor is sealed, the reaction proceeds, and the resulting products are extracted. Here, nothing enters or leaves the reactor while the reaction is occurring. However, this is not always the case.

Transport Rate—The transport rate of a substance is the rate which a substance moves into or out of a control volume.

One common type of reactor is a Continuously Stirred Tank Reactor (CSTR). This type of reactor contains a well-mixed volume. Typically, a CSTR will have a constant volume of liquid material. When a feed stream enters containing reactants, this forces a product stream to leave. The product stream contains whatever was in the well-mixed reactor.

In addition to flow in a pipe, there are other methods of species transport. A gas could evolve from a liquid and leave a reacting system. A permeable membrane such as a cell wall could allow for certain species to move from one side of the membrane to the other. A filtration medium could separate and remove solids from a liquid mixture. One should realize that the reaction rate and transport rate both may be responsible for bringing a species into a control volume or removing a species from a control volume. As a result, both the total reaction rate and the total transport rates should have matching units.

Given these fundamental concepts, one may now construct a model of a system.

Species Balance—A species balance can be written for each substance in a control volume. The species balance equations includes transport and reaction rates.

Given each molecular species in a control volume, a balance equation can be written. The general form of a species balance equation is in the form:

Accumulation rate = Σ Transport Rate In $-$ Σ Transport Rate Out

$+\Sigma$ Reaction Rate Creation $-$ Σ Reaction Rate Consumed

This means that for a single species in a control volume, the rate that species is accumulating is the volume is a summation of all the transport and reaction rates for that species. If the species is flowing into the control volume or being created by reaction in the control volume, the related terms are positive. If the species is being transported out of the volume or it is being consumed in a chemical reaction, the related terms are negative.

The one term not yet described in this balance equation is the accumulation rate.

Accumulation Rate—The rate at which a species changes in a control volume based on the total imbalance of transport and reaction rates.

The accumulation rate describes the total rate of change for a species in a given control volume. If a system is at steady-state, this means the accumulation rate is zero. Under the steady-state assumption, the balance equation simplifies to:

$0 =$ Σ Transport Rate In $-$ Σ Transport Rate Out

$+\Sigma$ Reaction Rate Creation $-$ Σ Reaction Rate Consumed

The steady-state assumption does not mean that the reaction rate is zero or that the transport rates are zero. A steady-state assumption means that the positive terms balance the negative terms in the species balance.

4.4 Mass Balance Examples

Example 4.2 Simple mixing tank (Fig. 4.1)

Consider a mixing tank. The mixing tank has an agitation system in place. Due to constant agitation, it is assumed that the tank is well-mixed. A solution containing species A flows into the tank. Thea volumetric flow rate of species A into the tank is F $\left(\frac{L}{s}\right)$. This feed flow contains species A at concentration $C_{Ao}(t)$ $\left(\frac{mol}{L}\right)$. Therefore, the total transport rate of species A into the tank in units of $\frac{mol}{s}$ is given by:

$$F\left(\frac{L}{s}\right) C_{Ao}(t) \left(\frac{mol}{L}\right) = F\, C_{Ao}(t)$$

This transport rate will have units of $\frac{mol}{s}$.

Assume that the tank has volume V (L) and the volume does not change with time. This means that as liquid flows into the tank other liquid flows out of the tank. We know that the flow rate into the tank is F $\left(\frac{L}{s}\right)$. This means that the volumetric flow rate out of the tank is F $\left(\frac{L}{s}\right)$. However, the concentration flowing out of the tank depends on the concentration in the tank. Assuming the tank is well-mixed, the concentration of species A in the tank would be $C_A(t)$ $\left(\frac{mol}{L}\right)$, a different value from the concentration flowing in, $C_{Ao}(t)$ $\left(\frac{mol}{L}\right)$. This means that the transport rate of species A out of the tank in units of $\frac{mol}{s}$ is:

$$F\left(\frac{L}{s}\right) C_A(t) \left(\frac{mol}{L}\right) = F\, C_A(t)$$

Fig. 4.1 A simple constant volume mixing tank with a single species

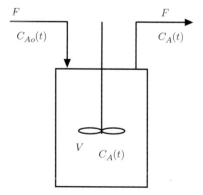

The overall balance on this mixing tank can be written as:

$$\text{accumulation rate} = (F \ C_{Ao}(t)) - (F \ C_{Ao}(t))$$

The accumulation rate in this case is derived from the rate of change of species A in the control volume. The total amount of A in volume V is

$$V \ (L) \ C_A(t) \left(\frac{\text{mol}}{L} \right) = V \ C_A(t)$$

so that the rate of change is determined by taking the time derivative.

$$\frac{d}{dt} \ (V \ C_A(t))$$

Since the volume is constant, this leads to:

$$V \ \frac{dC_A}{dt}(t)$$

Note that the overall units for this accumulation rate are units of $\frac{\text{mol}}{s}$ matching the units of the transport rates in and out of the mixing tank.

The species balance for A can now be written as:

$$V \ \frac{dC_A}{dt}(t) = F \ C_{Ao}(t) - F \ C_A(t)$$

At steady-state, the accumulation rate goes to zero. This makes sense as the time derivative would also go to zero. Assuming a steady-state feed concentration of C_{Aoss} and a steady-state product concentration of C_{Ass}, the steady-state mass balance becomes:

$$0 = F \ C_{Aoss} - F C_{Ass}$$

This implies that the outlet concentration will match the feed concentration at steady-state under the given assumptions.

Example 4.3 Continuous stirred tank reactor (Fig. 4.2)

Modeling a continuous stirred tank reactor is approximately as complex as modeling a simple mixing tank. Consider a reactor where species A reacts to create species B. This reaction does not instantaneously occur; the volumetric reaction rate is given by the expression:

$$r(t) = k \ C_A(t) \ \left(\frac{\text{mol}}{s L} \right)$$

Fig. 4.2 A simple constant volume stirred tank reactor schematic

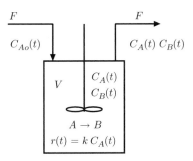

Since the volumetric reaction rate has overall units of $\left(\frac{\text{mol}}{\text{s L}}\right)$ and concentration is usually measured in $\left(\frac{\text{mol}}{\text{L}}\right)$, the reaction rate coefficient k must have units of $\left(\frac{1}{\text{s}}\right)$ or $\left(\text{s}^{-1}\right)$.

This reaction rate represents the rate at which species A is consumed by the reaction. Since the stoichiometric coefficients for the balanced reaction $A \rightarrow B$ are both 1, the reaction rate also represents the rate of production of species B. For every molecule of A consumed, a single molecule of B is generated.

The given reaction rate is volumetric. This means it is expressed per unit volume. To determine the total reaction rate in the reactor, one must multiply by volume:

$$V k C_A(t) \quad \left(\frac{\text{mol}}{\text{s}}\right)$$

Assume that species A flows into the reactor at concentration $C_{Ao}(t)$ and no species B flows into the reactor. Allowing the concentrations inside the reactor to be $C_A(t)$ and $C_B(t)$, the two balance equations can be written as:

$$V \frac{dC_A}{dt}(t) = F\,C_{Ao}(t) - F C_A(t) - V k\,C_A(t)$$

$$V \frac{dC_B}{dt}(t) = -F\,C_B(t) + V k\,C_A(t)$$

Note that the total reaction rate term appears in both model equations. The only difference being that the reaction term is negative when a species is being consumed and positive when a species is being produced (Fig. 4.2).

Example 4.4 Simple cell model (Fig. 4.3)

One may consider modeling of a single cell. To simplify the analysis, the cell may be considered a constant volume, well-mixed system. For the following cell model, there are three species: A, B, and C. The volumetric reaction rate for transforming A to B is given as:

$$r_1(t) = k_1\,C_A(t) \quad \left(\frac{\text{mol}}{\text{s L}}\right)$$

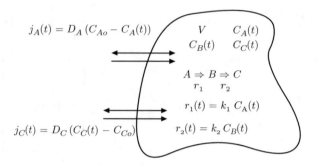

Fig. 4.3 A cell model with three species and two reactions

The volumetric reaction rate for the production of C from B is given as:

$$r_2(t) = k_2 \, C_B(t) \left(\frac{\text{mol}}{\text{s L}} \right)$$

This defines the reaction rates for the two reactions involving three species.

Molecules can be transported across the cell wall into and out of the cell control volume under consideration. Assuming that species A diffuses into the cell at a rate

$$j_A = D_A \, (C_{Ao} - C_A(t)) \left(\frac{\text{mol}}{\text{s}} \right)$$

where the transport rate $j_A(t)$ is given in units of $\left(\frac{\text{mol}}{\text{s}} \right)$. Here, D_A is the effective diffusion rate coefficient. The difference between the external concentration of A and the concentration of A in the cell is the driving force that transports species A into the cell.

Similarly, species C leaves the cell at the given rate:

$$j_C = D_C \, (C_C(t) - C_{Co}) \left(\frac{\text{mol}}{\text{s}} \right)$$

Now, the reaction rates and transport rates are defined. Three species balance equations can now be written as:

$$V \frac{dC_A}{dt}(t) = D_A \, (C_{Ao} - C_A(t)) - V \, k_1 \, C_A(t)$$

$$V \frac{dC_B}{dt}(t) = V \, k_1 \, C_A(t) - V \, k_2 \, C_B(t)$$

$$V \frac{dC_C}{dt}(t) = V \, k_2 \, C_B(t) - D_C \, (C_C(t) - C_{Co})$$

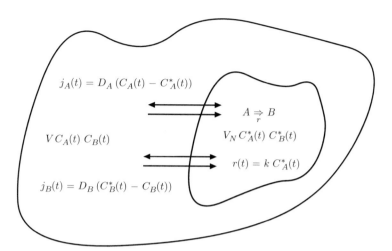

Fig. 4.4 A cell model with two species and two compartments

The three balance equations represent a simple cell model where two reactions are taking place inside the cell. Note that the reaction rates were given as volumetric rates. The total reaction rate is found by multiplying by the volume where the reaction is occurring. Also note the transport rates across the cell wall. In the odd case where the concentration of A outside the cell was very low while the concentration of A inside the cell was high, the transport rate term for A flowing into the cell would become negative. This case is also true for removal of C from the cell in the case where there is a high concentration of C outside the cell.

Example 4.5 Two compartment cell model (Fig. 4.4)

In this example, consider modeling of a cell where A moves into the nucleus of the cell where it is modified to species B. Species B then leaves the nucleus. The volume of the cell outside the nucleus is V and the effective volume inside the cell is V_N. The concentration of A and B in the cytoplasm outside the nucleus can be described by $C_A(t)$ and $C_B(t)$. The concentration of A and B inside the nucleus can be described as $C_A^*(t)$ and $C_B^*(t)$.

The total conversion rate for transforming A to B in the cell nucleus is:

$$r(t) = k\, C_A^*(t) \left(\frac{\text{mol}}{\text{s}} \right)$$

This reaction does not occur outside of the nucleus. The transport rate of A into the nucleus can be described as

$$j_A = D_A \left(C_A(t) - C_A^*(t) \right) \left(\frac{\text{mol}}{\text{s}} \right)$$

and the total transport rate of B out of the cell nucleus into the cytoplasm is:

$$j_B = D_B \left(C_B^*(t) - C_B(t) \right) \left(\frac{\text{mol}}{\text{s}} \right)$$

There is no appreciable transport across the cell wall.

There are now two control volumes, the volume of the cytoplasm and the volume of the cell nucleus. With two species in each volume, four total model equations can be used to describe the system.

$$V \frac{dC_A}{dt}(t) = -D_A \left(C_A(t) - C_A^*(t) \right)$$

$$V \frac{dC_B}{dt}(t) = D_B \left(C_B^*(t) - C_B(t) \right)$$

$$V_N \frac{dC_A^*}{dt}(t) = D_A \left(C_A(t) - C_A^*(t) \right) - k\, C_A^{/*}(t)$$

$$V_N \frac{dC_B^*}{dt}(t) = k\, C_A^{/*}(t) - D_B \left(C_B^*(t) - C_B(t) \right)$$

This model shows that when you react or transport a species, the rate term shows up negative in one balance and positive in another.

Problems

4.1 Consider the following problem to model oxygen carrying species and reaction inside a cell. Determine a model for the system based on the information provided. What other key assumption/s have you made? Assume the following:

- The oxygen carrying species in the cell is species B
- Species B has concentration $C_B(t)$ $\left(\frac{\text{mol}}{\text{L}} \right)$
- Species B diffuses into the cell
- The diffusion rate in is: $D_B A \left(C_{Bo}(t) - C_B(t) \right)$ $\left(\frac{\text{mol}}{\text{s}} \right)$
- Species B is converted to species C inside the cell
- The conversion occurs at a total molar rate of $V \frac{C_B(t)}{k + C_B(t)}$ $\left(\frac{\text{mol}}{\text{s}} \right)$
- Species C diffuses out of the cell
- The diffusion rate out is: $D_C A \left(C_C(t) - C_{Co}(t) \right)$ $\left(\frac{\text{mol}}{\text{s}} \right)$
- The effective cell volume is V (L)
- $C_B(t)$ is the molar concentration of B inside the cell
- $C_C(t)$ is the molar concentration of C inside the cell

- $C_{Bo}(t)$ is the molar concentration of B outside the cell
- $C_{Co}(t)$ are the molar concentration of C outside the cell
- Concentrations are assumed to be functions of time t

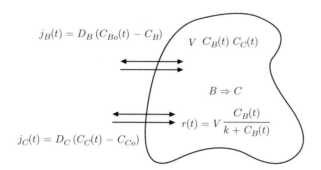

$$j_B(t) = D_B \left(C_{Bo}(t) - C_B \right)$$

$$V \quad C_B(t) \, C_C(t)$$

$$B \Rightarrow C$$

$$r(t) = V \frac{C_B(t)}{k + C_B(t)}$$

$$j_C(t) = D_C \left(C_C(t) - C_{Co} \right)$$

4.2 Consider ethanol metabolism in the body. Liquids are consumed, entering the gut. Ethanol transport into the blood stream can be modeled as a diffusion process. Ethanol is removed from the body due to urination and metabolism of ethanol in the blood. You must develop a dynamic model of this process relating the inlet concentration of ethanol to the resulting concentration in the blood. Assume the following:

- Liquid enters and leaves the body at a constant volumetric flow rate
- The volumetric flow rate is given by F (m^3/s)
- Entering liquid contains a time-varying concentration of ethanol
- The concentration of ethanol is given as $C_o(t)$ (mol/m^3)
- The gut can be assumed to be well-mixed volume V (m^3) at concentration $C_G(t)$ (mol/m^3)
- The blood stream can be assumed to be well-mixed with volume V_B (m^3) at concentration $C_B(t)$ (mol/m^3)
- The **total rate** of ethanol transferred from the gut to the blood is: $j(t) = DA(C_G(t) - C_B(t))$ (mol/s)
- The **total rate** of ethanol removed from the blood is $r(t) = kC_B(t)$ (mol/s)
- Physical properties and parameters do not change with time

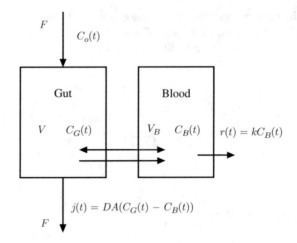

$$j(t) = DA(C_G(t) - C_B(t))$$

4.3 A continuous bioreactor for yeast fermentation has a single glucose feed stream. The growth reaction takes place in a simple CSTR. The inlet glucose concentration can be specified and adjusted. Develop a model for glucose in the system from the information provided. Assume the following:

- The reactor is well-mixed with constant volume V (L)
- The reactor glucose feed concentration is $C_{Gin}(t)$ $\left(\frac{\text{mol}}{\text{L}}\right)$
- The reactor glucose concentration is $C_G(t)$ $\left(\frac{\text{mol}}{\text{L}}\right)$
- The yeast fermentation reaction consumes glucose
- The reaction is a first-order reaction with **volumetric** reaction rate $k\,C_G(t)$ $\left(\frac{\text{mol}}{\text{L min}}\right)$
- The reactor volumetric flow rate in (and out) is F $\left(\frac{\text{L}}{\text{min}}\right)$

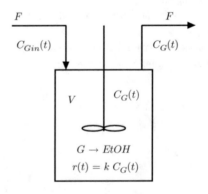

4.4 Consider a cell with three internal species. Develop a dynamic model for species A, B, and C in a cell using the following assumptions:

- The effective volume of the cell is V
- The cell may be assumed to be constant volume and well-mixed
- Species A enters the cell due to diffusion, where the total rate depends on the external concentration of A, $C_{Ao}(t)$
- The total diffusion rate of A into the cell is $j(t) = DA(C_{Ao}(t) - C_A(t))$ $\left(\frac{\text{mol}}{\text{min}}\right)$
- B and C do not diffuse in or out of the cell
- A reacts to form B with a total reaction rate of $r_1(t) = k_1 C_A(t)$ $\left(\frac{\text{mol}}{\text{min}}\right)$
- A also reacts to form C with a total reaction rate of $r_2(t) = k_2 (C_A(t))^2$ $\left(\frac{\text{mol}}{\text{min}}\right)$
- B can react to form C with a total reaction rate of $r_3(t) = k_3 C_B(t)$ $\left(\frac{\text{mol}}{\text{min}}\right)$
- C decomposes inside the cell with a total reaction rate of $r_4(t) = k_4 C_C(t)$ $\left(\frac{\text{mol}}{\text{min}}\right)$

4.5 A three-stage reactor system is used to purify a feed stream before the final product is produced. The reaction takes place in three different CSTRs to remove the contaminant species A. Develop a model to describe the concentration of A in the reactor system from the information provided. Assume the following:

- The reactors are well-mixed with constant volumes V_1 (L), V_2 (L), and V_3 (L)
- The feed concentration to reactor 1 is $C_{A0}(t)$ $\left(\frac{\text{mol}}{\text{L}}\right)$
- The reactor concentrations are $C_{A1}(t)$ $\left(\frac{\text{mol}}{\text{L}}\right)$, $C_{A2}(t)$ $\left(\frac{\text{mol}}{\text{L}}\right)$, and $C_{A3}(t)$ $\left(\frac{\text{mol}}{\text{L}}\right)$
- Species A is removed in each reactor
- The reactor volumetric flow rate in (and out) is F $\left(\frac{\text{L}}{\text{min}}\right)$
- Reactor 1 exhibits a **volumetric** reaction rate of $k_1 (C_{A1}(t))^2$ $\left(\frac{\text{mol}}{\text{L min}}\right)$
- Reactor 2 exhibits a **volumetric** reaction rate of $k_2 (C_{A2}(t))^2$ $\left(\frac{\text{mol}}{\text{L min}}\right)$
- Reactor 3 exhibits a **volumetric** reaction rate of $k_3 (C_{A3}(t))^2$ $\left(\frac{\text{mol}}{\text{L min}}\right)$

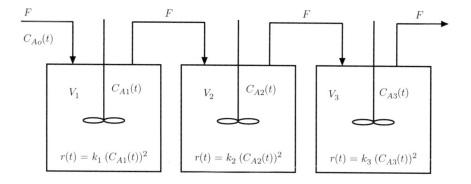

4.6 An aqueous pharmaceutical production process results in a side reaction making an undesirable component. A removal process has been proposed to remove the component from the product stream using a reactor and a settling tank. Develop a dynamic model for the following system with the following assumptions:

- The reactor is well-mixed with constant volume V (L)
- The mixing tank is well-mixed with constant volume V_M (L)
- The concentration of the contaminant fed to the reactor is $C_0(t)$ $\left(\frac{\text{mol}}{\text{L}}\right)$
- The concentration of the contaminant in the reactor is $C_1(t)$ $\left(\frac{\text{mol}}{\text{L}}\right)$
- The concentration of the contaminant in the mixing tank is $C_2(t)$ $\left(\frac{\text{mol}}{\text{L}}\right)$
- The removal reaction is a first-order reaction with **volumetric** reaction rate $k\,C_1(t)$ $\left(\frac{\text{mol}}{\text{L}\,\text{min}}\right)$
- There is **no** reaction in the mixing tank.
- There is no volume change upon mixing.
- The reactor volumetric flow rate in (and out) is F $\left(\frac{\text{L}}{\text{min}}\right)$
- The mixing tank volumetric flow rate in (and out) is F $\left(\frac{\text{L}}{\text{min}}\right)$

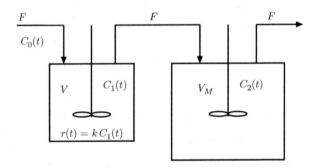

4.7 Consider a bioreactor system. Draw a simple diagram based on the following assumptions. Also, develop a dynamic model for the system based on these assumptions:

- There is a single rector of volume V (L)
- The reactor may be assumed to be constant volume and well-mixed
- Two species are in the reactor: A and B
- The concentration of A in the reactor is $C_A(t)$ $\left(\frac{\text{mol}}{\text{L}}\right)$
- The concentration of B in the reactor is $C_B(t)$ $\left(\frac{\text{mol}}{\text{L}}\right)$
- Dilute A and B enter the reactor at a volumetric flow rate of F $\left(\frac{\text{L}}{\text{min}}\right)$

- The concentration of A entering the reactor is given by $C_{A0}(t)$ $\left(\frac{\text{mol}}{\text{L}}\right)$
- The concentration of B entering the reactor is given by $C_{B0}(t)$ $\left(\frac{\text{mol}}{\text{L}}\right)$
- A single reaction takes place, converting species A into species B
- The volumetric reaction rate for converting A into B is given by $r(t) =$ $k_1 C_A(t)$ $\left(\frac{\text{mol}}{\text{L min}}\right)$
- There is no volume change with reaction

4.8 Consider a bioreactor system. Draw a simple diagram based on the following assumptions. Also, develop a dynamic model for the system based on these assumptions:

- There is a rector of volume V (L) in series with a settling tank of volume V_S (L)
- The vessels may be assumed to be constant volume and well-mixed
- Two species are in the reactor: A and B
- The concentration of A in the reactor is $C_{A1}(t)$ $\left(\frac{\text{mol}}{\text{L}}\right)$
- The concentration of B in the reactor is $C_{B1}(t)$ $\left(\frac{\text{mol}}{\text{L}}\right)$
- Dilute A enter the reactor at a volumetric flow rate of F $\left(\frac{\text{L}}{\text{min}}\right)$
- The concentration of A entering the reactor is given by $C_{A0}(t)$ $\left(\frac{\text{mol}}{\text{L}}\right)$
- No B enters the reactor
- A single reaction takes place, converting species A into species B
- The volumetric reaction rate for converting A into B is given by $r(t) =$ $k_1 C_A(t)$ $\left(\frac{\text{mol}}{\text{L min}}\right)$
- There is no volume change with reaction
- The flow out of the reactor into the settling tank is F $\left(\frac{\text{L}}{\text{min}}\right)$
- The concentration of A in the settling tank is $C_{A2}(t)$ $\left(\frac{\text{mol}}{\text{L}}\right)$
- The concentration of B in the settling tank is $C_{B2}(t)$ $\left(\frac{\text{mol}}{\text{L}}\right)$
- No reaction takes place in the settling tank
- The flow out of the settling tank is F $\left(\frac{\text{L}}{\text{min}}\right)$

4.9 Consider a bioreactor system. Draw a simple diagram based on the following assumptions. Also, develop a dynamic model for the system based on these assumptions:

- There are two rectors in series
- The volume of reactor 1 is V_1 (L) and the volume of reactor 2 is V_2 (L)
- The reactors may be assumed to be constant volume and well-mixed
- Three species are present in the reaction system: A, B, and C
- Dilute A enters the reactor at a volumetric flow rate of F $\left(\frac{\text{L}}{\text{min}}\right)$

- The concentration of A entering the reactor is given by $C_{A0}(t)$ $\left(\frac{mol}{L}\right)$
- There is no B or C in the feed stream to the first reactor
- The concentration of A in reactor 1 is $C_{A1}(t)$ $\left(\frac{mol}{L}\right)$
- The concentration of B in reactor 1 is $C_{B1}(t)$ $\left(\frac{mol}{L}\right)$ and there is no C in reactor 1
- A single reaction takes place in reactor 1, converting species A into species B
- The volumetric reaction rate for converting A into B is given by $r_1(t) = k_1 C_{A1}(t)$ $\left(\frac{mol}{L\,min}\right)$
- The flow from reactor 1 into reactor 2 is F $\left(\frac{L}{min}\right)$
- A single reaction takes place in reactor 2, converting species B into species C
- The volumetric reaction rate for converting B into C is given by $r_2(t) = k_2 C_{B2}(t)$ $\left(\frac{mol}{L\,min}\right)$
- There is no volume change with reaction
- The flow from reactor 2 is F $\left(\frac{L}{min}\right)$

4.10 Consider a cell with three internal species. Develop a dynamic model for species A, B, and C in a cell using the following assumptions:

- The effective volume of the cell is V
- The cell may be assumed to be constant volume and well-mixed
- The concentration of A inside the cell is given by $C_A(t)$ $\left(\frac{mol}{L}\right)$
- The concentration of B inside the cell is given by $C_B(t)$ $\left(\frac{mol}{L}\right)$
- The concentration of C inside the cell is given by $C_C(t)$ $\left(\frac{mol}{L}\right)$
- A, B, and C do not diffuse in or out of the cell
- A reacts to form B with a total reaction rate of $r_1(t) = k_1 C_A(t)$ $\left(\frac{mol}{min}\right)$
- A also reacts to form C with a total reaction rate of $r_2(t) = k_2 C_A(t)$ $\left(\frac{mol}{min}\right)$
- There is no volume change on reaction

4.11 An aqueous process requires multiple mixing stages to agitate a stream containing a single species of interest. Develop a dynamic model for the following system with the following assumptions.

- The first mixing tank is well-mixed with constant volume V_1 (L)
- The second mixing tank is well-mixed with constant volume V_2 (L)
- The concentration of the stream fed to the tanks is $C_0(t)$ $\left(\frac{mol}{L}\right)$
- The concentration of the species of interest in the first tank is $C_1(t)$ $\left(\frac{mol}{L}\right)$
- The concentration of the species of interest in the second tank is $C_2(t)$ $\left(\frac{mol}{L}\right)$
- There is **no** reaction in either mixing tank
- There is no volume change upon mixing

- The feed stream volumetric flow rate to **both** tanks is F $\left(\frac{L}{min}\right)$
- The first mixing tank outlet flows into the second tank at the rate: F $\left(\frac{L}{min}\right)$

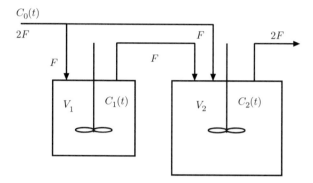

Chapter 5
Structured Programming

5.1 Introduction to Structured Programming

Engineers often use computers to solve problems. Sometimes engineers use a high level applications (programs) to perform engineering simulations. For mechanical systems, tools include AutoCAD, SOLIDWORKS, CATIA, and others for computer aided design. For electrical systems, pSPICE, AutoDesk Eagle, Fritzing, and others can be used for circuit simulation and design. For chemical processes, Hysys and Aspen are commonly used. Engineers should understand the fundamental building blocks that are used to develop these programs and application packages. In some cases, existing applications cannot do exactly what must be done and engineers must construct specialized novel methods to solve the given problem. In other cases, the software package can be extended to be much more useful if the engineer can write a small bit of code needed for the project. In still other cases, an engineer can use software code to operate software packages and analyze the resulting simulation data. The engineer must know engineering principles, computer language structures, and numerical methods to arrive at a solution. Computer programming concepts are one part of the solution.

Various computer languages and environments allow people to use very general concepts to come up with solutions to problems. These concepts are quite portable, in that the same concepts for structured problem solving exist in almost every language or environment. The following is a list of the basic concepts in this chapter

- Variables
- Assignment Statements
- Input and Output
- Basic Data Structures
- IF statements
- FOR statements

© Springer Nature Switzerland AG 2022
E. Gatzke, *Introduction to Modeling and Numerical Methods for Biomedical and Chemical Engineers*, https://doi.org/10.1007/978-3-030-76449-4_5

- WHILE statements
- Functions
- Debugging
- Pseudo Code
- Compiling vs. Interpreting

This chapter generally uses MATLAB syntax for expressing computational concepts. This should not cause substantial concern, as the same basic data types and operations are found in almost every programming environment. Variables are used to represent different data types and flow control is used with some variant of IF / FOR / WHILE. Knowledge of basic MATLAB syntax would facilitate a jump start into many other programming languages such as C, C++, C#, Java, VBA, .NET, Python, and many others. Most problems solved using MATLAB can also be solved using Mathcad or Excel.

5.2   Variables

Computers store information in memory. At a very low level, the operating system (Windows, Mac OS, Linux, Unix) keeps track of available memory and what application is using memory. When an application needs more memory, it asks for a certain amount of memory. If the memory is available, a memory address is provided to the program to use. Memory locations are typically referenced using hexadecimal notation, like FF01F2 or 00DA5F. Luckily, we do not usually need to keep track of memory addresses. Generally, we use **variables** to reference values in the memory. The actual hexadecimal address may change, but the variable continues to contain the same information.

Variable—A variable in a computer programming language is a human-readable label that represents data stored in computer memory.

Variables usually have names that are mostly characters. Some programming environments have limitations for variables names. For example, some programming environments are case sensitive (counter and COUNTER would represent different values). Some programming environments limit the length of variable names or limit the characters in a variable name.

It really is a personal decision for you to use whatever variable name you want to that is legal for the environment you are working in. The name should be descriptive enough so you have some idea what it is, but if it is too long you will have trouble typing it over and over and might make more mistakes. Most programming environments require that variable names start with a character, so "2counter" would not be valid. Also, spaces in variable names are not usually allowed, like "counter 2" or "Image 3". In many cases, the underline character is used for a space as in "counter_4" and "Image_7". Most special character should also be avoided in variable names.

The following are valid variable names in MATLAB:

- counter
- Counter
- new_counter
- NewCounter
- New_Counter
- Counter1
- counter_1

MATLAB does not require you to specify the type or size of variables before you use them. Some languages force you to specify exactly what type and size each variable is before it is used in a program.

5.3 Assignment Statements

You can assign a variable a new value using an assignment statement. The value on the right hand side of the equality is evaluated and assigned to the variable named on the left hand side. Sometimes this is just simple assignment, like:

```
a=1
b=2
```

Sometimes this is more complicated, like:

```
c=a*b+3
```

In other cases, one may use simple math functions provided by the environment:

```
a=sin(2)
b=log(3)
```

Note that you do not have to use just numbers in the function call, you can use variables in the functions:

```
c=sin(a)+log(b)
```

You can call a function using the current value of variable in the function call, reassigning the variable value:

```
a=log(a)
```

Sometimes you write your own function to do something special:

```
c=MyFunction(3)
```

Sometimes a function takes multiple input arguments:

```
c=conv(a,b)
a=rand(3,2)
```

In some cases, you can have multiple output arguments in a function, so two variables are assigned values after the function is evaluated:

```
[a,b]=find(c)
[rows,cols]=size(A)
```

5.4   Input and Output

There are various ways to get information into and out of the computer. For input, sometimes you type in values at the keyboard when prompted, sometimes you load a data file. There are more interesting ways to get input values as well. Audio input can be read using a mic. Video input can be read using a camera. Various environmental sensors (for temperature, pressure, or others) can be read using specialized Data AcQuisition (DAQ) devices.

The following MATLAB command would prompt the user for information and assign the value to the specified variable:

```
age=input('What is your age in years?')
```

The variable age would contain the user input value. Note that the user could type in a number or something not a number. For advanced programming, you may want to check to see that the user input actually is what you expect (a positive number between 0 and 120 in this case).

In many cases, you may have a data file that must be read. For simple numeric files with m columns and n rows of numeric data, MATLAB can use the load command.

```
data=load('Filename')
```

This would read the specified file and put the $n \times m$ data values into the variable data. For more complex file reading, you can use fopen, fread, fwrite. For dealing with character arrays, type "help strfun" to see what functions are available in MATLAB. Note that single quotes were used to specify the file name. MATLAB does not use a backward single quote (') for defining a character array.

For output, there are various ways to present results. One easy way is to write text to the computer screen:

```
disp('Something is wrong!')
```

This would just print the message on the MATLAB screen. The string of characters to be printed is defined using the single quotes. This could also be used for numerical warnings:

```
disp(['The reactor temp is ' num2str(temp) ])
```

This would actually use the current number in the variable temp to make the output warning message. The num2str function converts the numerical value of the variable temp into a string. MATLAB allows the resulting string of characters to be concatenated onto the other set string of characters, "The reactor temp is" because of the square brackets surrounding the two strings.

You can save data to a file:

```
save filename a b c
```

This would save the values of the variables a, b, and c into the specified file. You could load those variables again using:

```
load filename
```

and you would have a, b, and c back in the memory of MATLAB. Note that this file format is not human readable. To make a nice looking text file that you could open and look at in a text editor, use:

```
save filename a b c --ascii
```

Another type of output is visual. The plot command is very powerful in MATLAB.

```
plot([1 2 3],[4 5 6],'x')
```

There are even ways to play audio data, producing output in the speakers

```
sound(rand(1000,1))
```

Using a DAQ system, you can send output to an actuator like a valve or a motor.

5.5 Data Types

There are some basic data types available to use for solving problems. The basic types are integers, real numbers, characters, and Boolean values.

Integers Integers are integral values, like 0, 1, -2, etc. Languages usually limit the maximum and minimum integer values. MATLAB on some platforms is limited to approximately $\pm 10e300$.

Real Numbers Real numbers are numbers that are not integral, like 5.5 and -3.333. These are usually referred to as doubles, short for double precision. On 32-bit machines, a double precision value uses 64 0/1 bits to represent the value.

One bit is for the sign, some of the bits are for the exponent value, and the rest represent the binary value of the numeric portion (mantissa).

Characters Characters are basically anything on the keyboard. ASCII characters were originally limited to 128 characters using only seven bits. More modern Unicode (UTF-8, UTF-16, and UTF-32) expands possible characters to include basically every possible symbol by using more bits. Even Star Trek's Klingon language has Unicode characters.

Booleans Boolean values are either TRUE or FALSE. In MATLAB, a positive numeric value means TRUE and 0 or anything negative is false. Logical operators return 0 or 1, but 5 or 0.5 would be seen as true, just like -2 would be seen as false.

Arrays (Vectors and Matrices) Many times you have numerous data structures that should be put in a list or grid. Arrays are really more complicated data types comprised of lower level pieces of data such as integers, reals, or characters. One significant data structure used by engineers is the array (sometimes called a vector or matrix). Arrays are just indexed list of data values. A one-dimensional array is called a vector. A two-dimensional array is a matrix. Vectors are all technically matrices in MATLAB, so they could be considered matrices of size $N \times 1$ or $1 \times N$ (column or row vectors). Defining arrays in MATLAB can be accomplished with square braces, spaces, commas, and semicolons. Commas or spaces define new columns and semicolons define new rows as:

```
a=[5  6  7]
b=[5 ; 6 ; 7]
```

The previous example would create a row vector and a column vector in MATLAB. You can access individual elements of an array, so a(2) would ask for the second element of variable a.

```
c=a(2)
```

MATLAB also allows you to access multiple elements of an array:

```
c=a(2:3)
```

This would assign c as a vector with only the second and third elements of vector a.
    Matrices are two-dimensional arrays. Each element in a matrix has a row and a column index.

```
a=[ 4  5  6 ;  7  8  9]
```

This makes a matrix with two rows and three columns. You can access the row and column elements as a(row,col) so

```
a(2,3)
```

would access the second row, third column element. In this case, a value of 9.

You can use variables as the index when accessing elements:

```
a=[ 4 5 6 ; 7 8 9]
b=2
c=a(b,b)
```

Which would return the 2, 2 element of a, in this case 8. Note that MATLAB will complain if you try to access an element of an array that has not been assigned yet, like a(5,5) in this example.

Characters can be used in an array. An array of characters is usually called a string:

```
a='This is a string.'
b=a(1:7)
```

This would put the first seven r elements of a into the variable b, in this case the result would be the character array "This is".

MATLAB can also create N dimensional arrays. Consider a still image as a 2D array of pixels. However, each pixel is actually represented by three integer values for the red, green, and blue value displayed at each pixel. Each RGB color has a 2D matrix of values in the image that are "stacked" to create the image. To extend this concept to video one would stack a sequence of images in time. Consider the MATLAB code:

```
VID(3,4,2,6)=255
nd=ndims(VID)
size(VID,nd)
```

This sets the 4D array element in the (3, 4, 2, 6) position to 255. The MATLAB function ndims returns the number of dimensions for a variable. The function size gives the maximum size of a specified dimension in a given variable.

Complex Data Structures MATLAB also allows you to have more complex data structures, so that multiple pieces of data are associated with a single variable.

```
person.age=30
person.name='Tom'
person.phone=5551234
```

This means you could pass the data structure or object to a function with a single variable name as in processInputData(person) rather than the form processInput-Data(name, age, phone).

5.6   IF Statements

Note that for conditional IF statements, MATLAB does not recognize capital terms: IF, ELSE, ELSEIF, and END. MATLAB only recognizes lower case terms: if, else, elseif, and end. All caps terms are used here in the code examples to highlight the important terms in the conditional statements. Using "IF" instead of "if" would create a syntax error in current versions of MATLAB.

IF statements allow you to check a logical condition. If the logical condition is met, you do something. If the condition is not true, you may do something else. You may also have multiple logical conditions inside the IF statement.

```
IF (1<0)
    disp('Amazingly, 1 is less than 0')
    c=1
END
```

Notice the indentation for the IF statement. Indentation helps read your code, as all indented lines are more obviously part of the IF statement. You can put only one statement to execute, or as many as you want. The optional ELSE statement allows you to have a backup action that will execute only when the IF condition is not met.

```
IF (a<b)
    disp('Apparently, a is LESS than b')
ELSE
    disp('Apparently, a is NOT LESS than b')
END
```

You can put extra logical conditions in an IF statement as well. This allows for multiple conditions to be checked. The condition after the IF statement is checked. If the condition is not true, the condition after the ELSEIF is checked. If that is also not true, the code after the ELSE statement executes.

```
IF (a<b)
    disp('Apparently, a is LESS than b')
ELSEIF (a>b)
    disp('Apparently, a is GREATER than b')
ELSE
    disp('Apparently, a must be EQUAL to b')
END
```

If you have multiple IF and ELSEIF statements. If one condition is met, the later ELSEIF and ELSE code will not execute even the later ELSEIF condition is true.

```
IF (a<5)
    disp('Apparently, a is LESS than 5')
ELSEIF (a>2)
    disp('Apparently, a is GREATER than 2')
    disp('a is also NOT LESS than 5')
ELSEIF (a<7)
    disp('This will never execute!')
```

```
        disp('Either (a<5) or (a>2) is true!')
    ELSE
        disp('This will never execute!)
        disp('Either (a<5) or (a>2) is true!')
    END
```

You can have an IF statement inside an IF statement. This is called a nested statement. In the following example, the first statement must be true to ever execute the inner IF statement.

```
    IF (a<b)
        IF (a>0)
            disp('a is LESS than b and positive')
        ELSE
            disp('a is LESS than b and not positive')
        END
    ELSE
        disp('a is NOT LESS than b')
    END
```

You can also have more complicated logical statements using AND and OR operators. MATLAB uses the & character for AND. MATLAB uses | for OR. These can result in very complex expressions that may be difficult to sort out logically:

```
    IF ((a<b) & (a>0))
        disp('a is LESS than b AND positive')
    ELSEIF ((a>b) | (a<0))
        disp('a is greater than b OR less than zero')
    END
```

5.7 FOR Statements

Note that MATLAB does not recognize capital terms FOR and END. MATLAB only recognizes lower case terms for and end. All caps terms are used here in the code examples to highlight the important terms in the flow control statements.

If you want to do something a few times and you know how many times you want to do it, use a FOR statement.

```
    FOR i=1:3
        a(i)=i*i;
    END
```

This will make variable a into a vector [1 4 9]. The loop will iterate three times. The variable i will take a value of 1 the first time, assigning the first location of variable a the value on the right hand side. Basically, the FOR loop is executing the following three commands as the counter i increases from 1 to 2 to 3 in this case as:

```
a(1)=1*1;
a(2)=2*2;
a(3)=3*3;
```

You can use variables for the index as well. In this case, the variable a is assumed to be a one-dimensional array. The variable len takes on the value of the length of a. The FOR loop will iterate, going through every value in the vector array variable a, taking the numeric value in the ith storage spot of the vector, computing the square of the value, and putting that squared value into position i of the vector array variable named b.

```
len=length(a)
FOR i=1:len
    b(i)=a(i)^2;
END
```

Therefore, no matter how long the vector a is, this will put the elements of a squared into b. The variable b will be a vector the same length as the vector a (assuming that variable b had no value initially.

You can also have nested FOR loops as well.

```
[rows,cols]=size(a)
FOR rowval=1:rows
    FOR colval=1:cols
        b(rowval,colval)=a(rowval,colval)+5
    END
END
```

5.8 WHILE Statements

Note that MATLAB does not recognize capital terms WHILE and END. MATLAB only recognizes lower case terms while and end. All caps terms are used here in the code examples to highlight the important terms in the flow control statements.

WHILE statements execute until some condition is met. This means they could execute forever, if the condition is not ever met. These statements are useful if you do not know how many times you want something to execute.

```
sum=0
data=1
WHILE (data>0)
    data=input('Enter a positive number, or 0 to quit')
    sum=sum+data
```

END
```
        disp(['The resulting sum is ' num2str(data)])
```

This would keep prompting the user for another number, and add that number to sum. When the user enters 0 (or any negative number) the loop breaks out and continues on, displaying the result.

5.9 Scripts

In MATLAB, you can start up the MATLAB environment and start entering commands at the command prompt, > >. This is great for simple things, but for anything slightly complex you may want to save your commands. Type "edit" at the prompt to open the text editor. You can type a bunch of commands in that text editor and save it on the computer. Usually the file name ends with a .m extension. To run your commands from the text file, you have multiple options.

Assuming the file name has no spaces and starts with a character, you can just type the file name at the prompt and all the commands in the file would execute (until it hits some sort of error or finishes). This also assumes MATLAB is currently specified the directory where you saved your file. The "current directory" at the top right tells you where MATLAB thinks it is right now. Hit the "..." button to change the directory.

For a few lines of text, you can highlight the selected commands from the text editor and copy-paste the text into the MATLAB window. CTRL-C is copy and CTRL-V is paste. You can also highlight the selection, then right click on the text and select "evaluate selection."

5.10 Functions

Script files end with a .m extension and do something specified by the text in the file. In some cases, you may want to generalize a procedure to do something, like compute the mean of a vector. You can specify a function in a file with a .m extension similar to a script. The function must be saved in a file named procedurename.m. The first line of the file has a specific format. For example, the following would need to be saved in a file myfunction.m in the current directory of MATLAB. (Typing edit myfunction at the command prompt will open the MATLAB editor and create a new file myfunction.m)

```
        function mean = myfunction(x)
        n=length(x)
        sum=0
        for i=1:n
```

```
        sum=mean+x(i)
    end
    mean=sum/n
```

This input function takes whatever input x, figures out the length of x, figures out the sum of the elements in the input, then calculates the mean of the vector. This assumes x is a vector (or a scalar value). You can call this function from another function or call it from a script or call it from the MATLAB prompt.

The input variable x is whatever you call the function with, so the following examples would work:

```
    myfunction( [ 1 2 3 4 ] )
    a = [ 3 4 5 6]
    myfunction(a)
```

Note that the variables n, sum and i are used inside the function. These are called local variables. If you had a variable named sum, n or i outside of the function, calling the function would not change the value of the variables outside the function.

```
    n=5
    myfunction( [ 1 2 3 ] )
```

Here, n would still be 5, although inside myfunction n will have a value of 3. You can specify multiple outputs for your functions as well.

```
    function [minval, maxval]=minmaxfunction(x)
    minval=x(1)
    maxval=x(1)
    n=length(x)
    for i=2:n
        if (x(i)<minval)
            minval=x(i)
        end
        if (x(i)>maxval)
            maxval=x(i)
        end
    end
```

Think about what happens in this function. If x is length 1, the resulting minval and maxval are just x. Otherwise, it goes through the indices of x from 2 to the end looking for a bigger and bigger or smaller and smaller values.

5.11 Naming Conventions

There are a variety of ways to pick names for variables and function names. In MATLAB, names must follow a few rules. Names can only include characters, numbers, and the underscore character. Names must start with a letter (not a number) and are case sensitive. Variables k and K would be considered different items. Most languages and environments have similar rules.

In MATLAB, 2D matrices are often named as upper case values like A or K. Of course, matrices can have more descriptive names like `Ainv` or `Kapprox`.

Some people prefer to only use lower case values so that there is no upper case/lower case confusion, like `aiv` or `kapprox`.

Many computer programmers use a convention where words are joined but the first letter of the words is capitalized. This is commonly referred to as *camel case* or *camelCase*. Some examples would be `newMatrixInverse` or `approximateGainMatrix`. *Capital Camel Case* would capitalize the leading letter as `NewMatrixInverse` or `ApproximateGainMatrix`.

Another option is commonly referred to as *snake case*. Here, underscores are used to separate terms: `new_matrix_inverse` or `approx_gain_matrix`.

In general, variable names should be descriptive. However, if the variable is too long the code can be difficult to read. Using comments with code can also help with making code understandable. Here are some pointers to avoid:

- Using the number 1 and the letter l can lead to confusion.
- Using numerous simple variables (x1, x2, x3, etc.) without description.
- Using lazy names (x, xx, xxx, xxxx) will lead to confusion.
- Not including any comments makes code impossible to read.

5.12 Debugging

Debugging just means fixing your code. For example, you may write code that goes past the end of a vector or does not produce the desired output. You must think about the variable values at each step in your code and think about what is happening at each step. This is called a variable trace. Sometimes it helps to print out variable values at many points in your code and see where things go wrong. MATLAB will print out variable values if you type a variable name by itself.

```
> > a=2
a =
     2
> > a
a =
     2
```

You can suppress this normal output by using a semicolon at the end of a line.

```
function out=myfunction(x)
count=0
out=0
while (count<length(x))
    out=x(count)
    count=count+1
end
```

This function will not work. The count variable starts at 0, so MATLAB will complain when you try to access the 0th element of the value x.

5.12.1 Pseudo Code

Pseudo Code is just a way of sketching out a solution methodology. Using pseudo code, you do not have to use accurate code syntax. You can summarize steps into a single idea, like "find the minimum and maximum values of the data" or "save the data in the specified format." See the flow charts of Chap. 1 for examples of a graphical process representation.

5.13 Compiling vs. Interpreting

On a PC, you have executable programs. These are special files with the commands needed to run something on the computer. Executable files on a PC end in .exe. These are created by compilers. You take source code written in C or C++ or Fortran or some other language and run it through a compiler to make an executable file. For example, MS Word is a .exe file and so is MATLAB. MATLAB, MathCAD, and Java are all interpreted environments. Interpreted files rely on some executable to be running. MATLAB figures out what to do for a given .m file. Interpreted environments are usually slower than compiled code.

5.14 Examples

These examples are provided using MATLAB syntax. The basic concepts are very similar for many other languages, like Python, C++, and Java.

Example 5.1 Basic Statements

Given the following MATLAB code, determine the final value for variable x after the code executes.

```
x=4
x=3*x+1
x=[x    2*x    0]
```

This example has three commands. The first command is x=4 which sets variable x to have a value of the number 4. Whatever variable x was before this line executes is forgotten. Going forward, variable x will have the value 4. Anywhere variable x appears could be replaced by the number 4.

The second line is given as: x=3*x+1. In most languages, the right hand side is evaluated and "pushed" into the variable on the left hand side. So, in this example we know from the first line that x has a value of 4. This means that the right hand side can be evaluated by sticking in a value of 4 wherever variable x appears. This results in the right hand side being 3*(4)+1 which is 13. This means after the second line executes, variable x has a value of 13.

The third line is a bit tricky: x=[x   2*x   0]. Square brackets in MATLAB can be used to create vectors and matrices. In this case, the right hand side has three things inside the square brackets, [x   2*x   0]. Since we know variable x has the value of 13, it can be replaced in every instance: [(13)   2*(13)   0] such that the right hand side is effectively [13   26   0]. This means that after these three lines execute, variable x will be a row vector of size 3 containing [13   26   0]. Alternatively, in MATLAB, this could also be described as a 1x3 matrix.

Note that MATLAB allows spaces or commas to separate columns in a matrix. Semicolons can be used to make new rows. Therefore the line of code x=[x ; 2*x ; 0] would create a 3x1 matrix or a column vector of size 3 that would be equivalent to:

$$\begin{bmatrix} 13 \\ 26 \\ 0 \end{bmatrix}$$

Semicolons are important when defining vectors and matrices. They are not so important when they appear at the end of a line to suppress screen output. However, consider the code:

```
a=rand(1000,1000)
```

This creates a 1000x1000 random matrix to be stored in variable a. Without a semicolon at the end of this line, all one million random elements of the new matrix a would be shown on the screen!

Example 5.2 IF Statement

Given the following MATLAB code, determine the final value for variable x after the code executes.

```
x=4^-1;
if (x<1)
   x=x*2;
else
   x=-x;
end
```

Here, the first line is a simple assignment statement: x=4^-1; which means variable x is set to a numeric value, 4^{-1} or $\frac{1}{4^1} = \frac{1}{4}$. After this line executes, the variable x will have a value of 1/4 or 0.25. Note that the semicolon at the end of the line of code suppresses output to the screen.

Now the condition in the IF statement is checked, (x<1). Since variable x has a value of 0.25, this condition will evaluate to true. This means the line x=x*2 will execute. This means variable x will now contain a value of 1/2 or 0.5.

In this example, the IF condition will always be true. This means the code after the ELSE statement will never execute. Also, all the assignment commands are followed by semicolons. This means the user will not see any values change on the screen but variable x will have a value of 0.5 after this section of code executes.

Example 5.3 Complex IF Statement

```
x=5
if (x>0)
     if (x<1)
          x=2*x
     elseif (x<3)
          x=4*x
     else
          x=5*x
     end
else
     x=6*x
end
```

This is a more complicated IF statement. It is basically an IF statement inside an IF statement. Consider the outer IF statement:

```
x=5
if (x>0)
     % CODE BLOCK A
else
     % CODE BLOCK B
end
```

The Boolean condition will be checked and either block A or block B will execute. Since variable x is assigned a value of 5, the condition will be true and block A will execute. Code block A is:

```
if (x<1)
        x=2*x
elseif (x<3)
        x=4*x
else
        x=5*x
end
```

Note how different things line up using proper indentation. Now, this IF statement consists of an IF condition, an ELSEIF condition, and an ELSE statement. This means the IF condition will be checked. Whenever that condition is not true, the ELSEIF condition will be checked. Whenever both the IF condition and ELSEIF conditions are not true, the ELSE condition code will execute.

In this case, variable x was assigned a value of 5 before the IF statements. Variable x has not changed, so it still has a value of 5. In this block, the condition (x<1) is not true so the statement x=2*x will not execute. The ELSEIF condition (x<3) is also not true, so the statement x=4*x will not execute. Finally, the else condition catches anything that gets past the IF and ELSEIF conditions so the line x=5*x will execute. This changes the variable x to a value of 25.

Example 5.4 **FOR Loop**

Given the following MATLAB code, determine the final value for variable x after the code executes.

```
x=[ ]
for k=1:3
        x(1,k)=2*k
end
```

The first line is a way of clearing variable x. It creates an empty 0x0 matrix for variable x. This is a way to make sure variable x does not contain any other values before the rest of the code starts. Another way to achieve this in MATLAB would be to use the line of code:

```
clear x
```

In this example code, the FOR loop line uses a counter variable k. The variable k is assigned an initial value of 1. The loop will iterate through with k having a value of 1, then k having a value of 2, then k having a value of 3.

When variable k has a value of 1, the code in the FOR loop x(1,k)=2*k evaluates to x(1,1)=2*1 which puts the value 2 in the 1,1 position of variable x. At this point, variable x is just a scalar since the "size" is 1x1.

When variable k has a value of 2, the code in the FOR loop x(1,k)=2*k evaluates to x(1,2)=2*2 which puts the value 4 in the 1,2 position of variable x. This means x is now a 1x2 matrix (row vector) containing the values [2 4]. In MATLAB, you can typically add elements to a vector or matrix without specifying the size of the variable. Other languages often require you to know how large a matrix or variable is before you can use that variable.

When variable k has a value of 3, the code in the FOR loop x(1,k)=2*k evaluates to x(1,3)=2*3 which puts the value 6 in the 1,3 position of variable x. This means variable x is now a 1x3 matrix (row vector) containing the values [2 4 6].

The same effect could have been accomplished using the following three lines of MATLAB code:

```
x(1,1)=2*1
x(1,2)=2*2
x(1,3)=2*3
```

However, what if the for loop final limit were 300 instead of 3? You would not want to copy three hundred lines! Whenever something must be done repeatedly and the number of iterations required is known, a FOR loop should be used.

Note that the default increment is 1 in a FOR loop. Something like:

```
for k=1:3:11
```

would evaluate the code inside the FOR loop with counter variable k having a value of 1, then 4, then 7, then 10. In the next iteration, counter variable k would be 13 which exceeds the limit of 11 so it would not execute again.

Example 5.5 WHILE Loop

Given the following MATLAB code, determine the final value for variable x after the code executes.

```
x=10
while x>1
    x=x/2
end
```

Here, variable x starts with a value of 10 when starting the WHILE loop. In this case, the condition (x>1) is checked. Since variable x has a value of 10, the condition is true and the code inside the WHILE loop will execute.

The first time the code executes, variable x has a value of 10 so x=x/2 will change the value of variable x from 10 to 5. After the code inside the WHILE loop is done, MATLAB returns to the top of the loop to start again.

Now, the condition (x>1) is checked with variable x having a value of 5. This condition is true again, so the loop executes. The statement x=x/2 evaluates, changing the value of variable x from 5 to a value of 2.5. The next iteration, variable x will change to 1.25.

The last iteration, variable x has a value of 1.25. This value is still greater than 1, so the code executes one final time. Variable x changes from 1.25 to 0.675. The next time the WHILE condition is checked, variable x will not be greater than 1. This means the condition is not true and the WHILE loop is complete. Variable x will retain the value it had, 0.675.

Example 5.6 Complex Code

Given the following MATLAB code, determine the final value for variable x after the code executes.

```
x=[ ]
for i=1:3
    for k=1:i
        if (i*k+i>3)
            x(i,k)=10*i+k
        end
    end
end
```

Here, there is an IF statement inside a FOR statement inside another FOR statement. The first FOR loop has a counter variable i that ranges from 1 to 3 changing by 1 each time. This means variable i will have a value of 1, then 2, then 3. This is the outer loop. All the code inside this loop will execute three times.

The inner FOR loop has a counter variable k that will range from 1 to the current value of i. This means the inner FOR loop with variable k will have k go from 1 to 1 when i is 1, then k will go from 1 to 2 when i is 2, then k will go from 1 to 3 when i is 3.

This means that the IF statement will execute a total of six times, each time with a different value of i and k. The six times are:

```
i=1,  k=1
i=2,  k=1
i=2,  k=2
i=3,  k=1
i=3,  k=2
i=3,  k=3
```

These six cases will be checked in the IF condition (i*k+i>3)

```
i=1,  k=1,  1*1+1,  false
i=2,  k=1,  2*1+2,  true
i=2,  k=2,  2*2+2,  true
i=3,  k=1,  3*1+3,  true
i=3,  k=2,  3*2+3,  true
i=3,  k=3,  3*3+3,  true
```

Each time the condition is true, the code x(i,k)=10*i+k will execute. This will have the five following values of variable i and k execute:

```
i=2,  k=1,  x(2,1)=10*2+1
i=2,  k=2,  x(2,2)=10*2+2
i=3,  k=1,  x(3,1)=10*3+1
i=3,  k=2,  x(3,2)=10*3+2
i=3,  k=3,  x(3,3)=10*3+3
```

This means there will be a value of 21 in the 2,1 position, 22 in the 2,2 position, 31 in the 3,1 position, 32 in the 3,2 position, and 33 in the 3,3 position. MATLAB will put zeros in the other positions. The variable x will then be a 3x3 matrix with the values:

$$\begin{bmatrix} 0 & 0 & 0 \\ 21 & 22 & 0 \\ 31 & 32 & 33 \end{bmatrix}$$

Problems

5.1 What is the value of x after the following MATLAB code executes?
```
x=5
x=2*x+3
```

5.2 What is the value of y after the following MATLAB code executes?
```
y=4
y=y^2
y=y+8
```

5.3 What is the value of z after the following MATLAB code executes?
```
z=1/4
z=32*z*z
```

5.4 What is the value of k after the following MATLAB code executes?
```
k=1
k=k+2*k
```

5.5 What is the value of m after the following MATLAB code executes?
```
m=((4+6)/2)^2
```

5.6 What is the value of x after the following MATLAB code executes?
```
x=-2
x=x*x+3*x
x=x/2
```

5.7 What is the value of y after the following MATLAB code executes?
```
y=-4
y=y^3
y=-30+y
```

5.8 What is the value of z after the following MATLAB code executes?
```
z=1/2
z=16*z*z
```

5.9 What is the value of k after the following MATLAB code executes?
```
k=1
k=k+2*k
```

5.10 What is the value of *m* after the following MATLAB code executes?
```
m=((4+6)/2)^2
```

5.11 What is the value of *A* after the following MATLAB code executes?
```
A=[1 2 ; 3 4]
A(1,2)=A(1,2)+7
```

5.12 What is the value of *B* after the following MATLAB code executes?
```
B=[7 ; 5 ; 2]
B(2)=B(2)*2
```

5.13 What is the value of *C* after the following MATLAB code executes?
```
C=[ 2 3 4 ; 5 6 7]
C(:,2)=[9 ; 8]
```

5.14 What is the value of *D* after the following MATLAB code executes?
```
D=[2 3 4 5 6]
D(4:end)=[11 12]
```

5.15 What is the value of *E* after the following MATLAB code executes?
```
E=[2:2:8]
E=[E 10 12]
```

5.16 What is the value of *A* after the following MATLAB code executes?
```
A=ones(2,3)
A(1:2,2)=[3; 4]
```

5.17 What is the value of *B* after the following MATLAB code executes?
```
B=zeros(3,3)
B(2:end,3)=B(2:end,3)+2
```

5.18 What is the value of *C* after the following MATLAB code executes?
```
C=[ones(2,3) ; [ 2 3 4] ]
C(:,2)=[-1 ; -2 ; -3]
```

5.19 What is the value of *D* after the following MATLAB code executes?
```
D=[0:.1:1]
D=D+1
```

5.20 What is the value of *E* after the following MATLAB code executes?
```
E=[3:2:12]
E=[E 14 15]
```

5.21 What is the value of *a* after the following MATLAB code executes?
```
a=6
if (a<0)
    a=2*a
else
    a=a/2
end
```

5.22 What is the value of b after the following MATLAB code executes?
```
b=9
if ( (b>0) & (b<10) )
      b=b*10
else
      b=0
end
```

5.23 What is the value of c after the following MATLAB code executes?
```
c=4
if (c>10)
      c=c*10
elseif (c>5)
      c=c*5
else
      c=0
end
```

5.24 What is the value of x after the following MATLAB code executes?
```
x=-1
if (x>x*x)
      x=2*x;
else
      x=x/2;
end
```

5.25 What is the value of y after the following MATLAB code executes?
```
y=5
if (y<0)
      y=y+1;
end
```

5.26 What is the value of z after the following MATLAB code executes?
```
z=sort(5+4)
if (z<4)
      z=z+18
else
      z=-1
end
```

5.27 What is the value of a after the following MATLAB code executes?
```
a=-1
for i=1:3
      a=a+a;
end
```

5.28 What is the value of *b* after the following MATLAB code executes?

```
b=2
for i=1:4
        b=b*2
end
```

5.29 What is the value of *x* after the following MATLAB code executes?

```
x=[]
for i=1:4
        x(i)=i^2
end
```

5.30 What is the value of *y* after the following MATLAB code executes?

```
y=[]
for k=1:3
        y(k)=4^k
end
```

5.31 What is the value of *a* after the following MATLAB code executes?

```
a=3
while (a<30)
        a=a*a;
end
```

5.32 What is the value of *b* after the following MATLAB code executes?

```
b=1/2
while (b>.01)
        b=b/2
end
```

5.33 What is the value of *c* after the following MATLAB code executes?

```
c=1
while c<1000
        c=c*2
end
```

5.34 What is the value of *x* after the following MATLAB code executes?

```
x=4
if (x>0)
        if (x<10)
                x=99;
        else
        x=88;
        end
else
        x=77;
end
```

5.35 What is the value of *y* after the following MATLAB code executes?
```
y=-10
if (y>10)
        y=2*y;
elseif (y>0)
        y=3*y;
else
        if (y<-4)
                y=4*y;
        else
                y=5*y;
        end
end
```

5.36 What is the value of *a* after the following MATLAB code executes?
```
a=[]
for row=1:2
        for col=1:3
                a(row,col)=10-row-col;
        end
end
```

5.37 What is the value of *b* after the following MATLAB code executes?
```
b=[]
for k=1:4
        for m=1:5
                b(k,m)=k-m;
        end
end
```

5.38 What is the value of *c* after the following MATLAB code executes?
```
c=zeros(3,4)
[rows,cols]=size(c)
for row=1:rows
        for col=1:cols
                c(row,col)=row+col;
        end
end
```

5.39 What is the value of *d* after the following MATLAB code executes?
```
d=[]
for row=1:3
        for col=1:3
                d(row,col)=10-row+col;
        end
end
```

5.40 What is the value of *e* after the following MATLAB code executes?
```
e=ones(2,3)
[rows,cols]=size(e)
for row=1:rows
      for col=1:cols
            e(row,col)=3*row-col;
      end
end
```

Further Reading

1. Attaway, S. (2018). *MATLAB: A practical introduction to programming and problem solving* (5th ed.). Butterworth-Heinemann.
2. Axler. S. (2015). *Linear algebra done right* (3rd ed.). Springer.
3. Chapman, S.J. (2019). *MATLAB programming for engineers* (6th ed.). Cengage Learning.
4. Hahn, B., & Valentine, D.T. (2019). *Essential MATLAB for engineers and scientists* (7th ed.). Academic Press.
5. Larson, R. (2016). *Elementary linear algebra* (8th ed.). Cengage Learning.
6. Lay, D.C., Lay, S.R., & McDonald, J.J. (2014). *Linear algebra and its applications* (5th ed.). Pearson.
7. Moin, P. (2010). *Fundamentals of engineering numerical analysis* (2nd ed.). Cambridge University Press.
8. Moore, H. (2007). *MATLAB for engineers* (5th ed.). Pearson.
9. Strang, G. (1993). *Introduction to linear algebra*. Wellesley, MA: Wellesley-Cambridge Press.
10. Young, T. (2017). *Matlab programming for engineers*. Samurai Media Limited.

Chapter 6
Introduction to MATLAB

6.1 MATLAB Advantages and Disadvantages

MATLAB has many toolboxes that can increase the capability of the application. The MATLAB user community is fairly sizable. MATLAB is cross-platform, meaning that it can run on Windows PCs, desktop Macs, and Linux machines. MATLAB files are simple text files that can be read or modified by any text editor. MATLAB can be used on an iPad or Android device by connecting to an active desktop MATLAB installation. MATLAB can be relatively slow for large complex numerical problems. MATLAB does not require the user to define variable types before the variables are used. MATLAB is sometimes considered expensive for professional use. OCTAVE is a MATLAB-compatible open-source software project which is free for anyone to use (Figs. 6.1 and 6.2).

6.2 Commands

Mastering the following commands will provide a basic level ability in MATLAB. Use the help functionality to find more information on these commands, including syntax examples.

- defining vectors and matrices with []
- accessing elements of vectors and matrices
- accessing portions of vectors and matrices
- whos—to determine information about the variables in memory
- clear—used to remove variables from memory
- pwd—to display the current working directory where files are loaded or saved
- ones—to make a matrix full of ones
- zeros—to make a matrix full of zeros

© Springer Nature Switzerland AG 2022   99
E. Gatzke, *Introduction to Modeling and Numerical Methods for Biomedical and Chemical Engineers*, https://doi.org/10.1007/978-3-030-76449-4_6

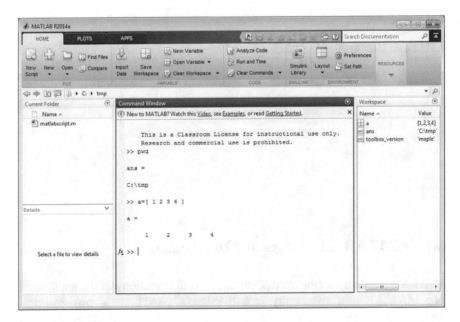

Fig. 6.1 MATLAB interface. The command prompt is in the central region. MATLAB commands can be entered to the right of the >> prompt. The left side of the display shows the current working directory where files are saved/loaded and the contents of the current directory. The right side of the display shows information about all variables in the MATLAB workspace

- end operator—accessing everything to the end of the matrix
- length—returns the length of a vector
- size—returns the size of a matrix or a vector
- if—conditional operator
- elseif—additional condition in IF statement
- else—alternate operations when IF and ELSEIF are not satisfied
- for—looping syntax when the number or iterations is known
- while—looping syntax that checks a condition at each iteration
- disp—function to print information to the command window
- num2str—function to convert a number into a string value for display
- input—reads data from the user
- inv—determines the inverse of a square matrix
- det—returns the determinant of a matrix
- eye—generates an identity matrix of a given size
- transpose operator—transposes a matrix
- anonymous functions—allows for definition of simple numerical functions
- fzero—finds values that make a given function have a value of 0
- roots—finds the roots of a polynomial
- poly—returns the value of a polynomial at a given value
- figure—creates a figure window for plotting data

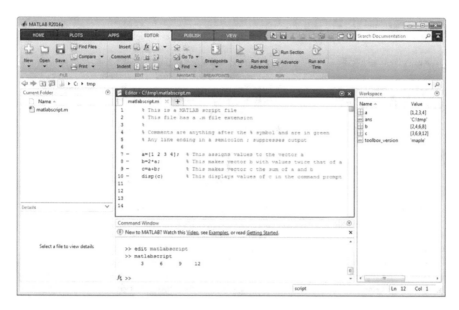

Fig. 6.2 MATLAB interface showing the editor window docked in the central area. Note that anything after the % symbol is a comment and will not execute. Comments are shown in green

- plot—creates an axis for plotting data
- clf—clears the current plot
- hold—allows you to plot over a plot
- subplot—used to put multiple plots on a single figure
- legend—used to label multiple sets of data
- xlabel—used to label the x axis
- ylabel—used to label the y axis
- title—used to give a title to a plot

6.2.1 Examples of Basic Matlab Commands

These are basic examples of Matlab commands that are quite useful. You should understand how the commands work, not just memorize these lines. To accomplish something you may have to modify or combine these commands.

```
clear                    % clear the memory
disp('This text shows')  % display a character array
% Following constructs a 3x3 matrix using [ ] and ;
A=[ 1 2 3 ; 4 5 6 ; 7 8 9]
b=[1 ; 2 ;3]             % makes a column vector
b=[1 2 3]               % makes a row vector
x=[0:.1:3]              % vector from 0 to 3 spaced by 0.1
b=b'                    % transpose a vector or matrix
```

```
c(1)=1;c(2:3)=[5 7]       % put elements into a vector
inv(A)                    % get the inverse matrix
xsol=inv(A)*b             % set xsol to solution of Ax=b
A*b                       % matrix vector multiplication
b.*b                      % element-by-element mult
[rows,cols]=size(A)       % get the size of a matrix
rows=size(A,1)            % size of a matrix in dimension 1
n=length(x)               % get length of a vector
eye(3)                    % create a 3x3 identity matrix
zeros(4,5)                % make a zero vector
cross(b,c)                % cross product of two vectors
dot(b,c)                  % dot product of two vectors
det(A)                    % determinant of a square matrix
A(2,3)                    % access portions of a matrix
A(2:3,1:2)                % rows 2 to 3, cols 1 to 2 of A
A(2:3,1:2)=zeros(2,2)     % Set part of matrix A to all 0s
A(2:end,3:end)            % row 2 to end and col 3 to end
A(1,:)                    % row 1, all columns of A
A(1,:)=ones(1,3)          % set row 1 of A to all ones
A(:,2:3)                  % all rows and cols 2 to 3 of A
f=@(x) x*x-2*x-1          % create an anonymous function f
fzero(f,1)                % find a root of function f starting at 1
roots([3 2 1])            % roots of polynomial 3x^2+2x+1=0
p=[2 0 1 0]               % make a vector p
roots(p)                  % find roots of 2x^3+0x^2+1x+0=0
polyval(p,x)              % value of polynomial p at points x
whos                      % show all variables and their size
pwd                       % print the working directory
```

6.2.2   Examples of Basic Matlab Plotting Commands

These Matlab commands can be useful for plotting.

```
clear                          % clear the memory
t=[0:1:40];                    % make vector for t
u=9*(t>1);                     % make a vector for u
y1=19.2*(1-exp(-.2*t));        % make y1 depend on t
y2=17.2*(1-exp(-.3*t));        % make y2 depend on t
plot(t,y1,'rx',t,y2,'bo')      % plot y1 and y2 vs t
% Note y1 will be red xs, y2 will be blue os
legend('Test 1','Test 2')      % add a legend
ylabel('Concentration \frac{mol}/{L}') % Add yaxis
xlabel('Time (s)')             % add xlabels
subplot(2,1,1)  % subplot figure with 2 rows, 1 col
plot(t,y1,'rx',t,y2,'bo')
legend('Test 1','Test 2')
ylabel('Concentration \frac{mol}/{L}')
subplot(2,1,2)                 % move to second subplot
plot(t,u)
```

```
ylabel('Temperature (C)')
xlabel('Time (s)')
data=50+10*randn(100,1);      % make data with avg 50,
                              % and std of 10
bins=[0:2:100];               % set bin size
subplot(1,1,1)
hist(data,bins)               % Make a histogra
M=randn(20,20)                % make 20x20 random matrix
mesh(M) %plot in 3D
```

6.3 Tutorials

The tutorials introduced here include links to online PDF files for the actual tutorial. This is intended so that tutorials can be updated when software changes necessitate modifications. The tutorials are quite lengthy and include step-by-step information suitable for a beginner. It is intended that students download the PDF file and start MATLAB. Students can split the screen with the PDF tutorial on one side of the screen and MATLAB on the other. Alternatively, students can switch between the PDF file tutorial and a full-screen MATLAB application. To rapidly switch screens on a Windows PC, use Alt-Tab.

6.3.1 MATLAB Basics

This tutorial introduces the MATLAB working environment, including the command prompt and the script editor. Variables of different types are assigned and investigated, including vectors and matrices. Students create different data types and learn how to access data elements. MATLAB commands introduced include whos, clear, pwd, ones, zeros, the end operator, length, and size. The tutorial leads the creation of a MATLAB script file that generates a single page text file with a snapshot of the MATLAB variable workspace. The PDF file can be downloaded from

http://www.cse.sc.edu/~gatzke/211/matlab1.pdf

6.3.2 MATLAB Flow Control

This tutorial introduces flow control syntax. IF statements, FOR loops, WHILE loops are considered. Nested structures are also presented. MATLAB commands introduced include if, elseif, else, for, while, disp, num2str, and input. The deliverable is a single page text file that summarizes the results of variables used to demonstrate IF, FOR, and WHILE commands in MATLAB. The PDF file can be downloaded from

http://www.cse.sc.edu/~gatzke/211/matlab2.pdf

6.3.3   MATLAB Linear Algebra and Root Finding

This tutorial considers concepts from linear algebra, including the matrix inverse and determinant of a matrix. Additionally, root finding procedures are presented. MATLAB commands introduced include dot, cross, inv, det, eye, the transpose operator, inline, fzero, poly, and roots. The deliverable is a MATLAB figure plotting a polynomial with text included on the figure to show the values of multiple variables used in. The PDF file can be downloaded from

http://www.cse.sc.edu/~gatzke/211/matlab3.pdf

6.3.4   MATLAB Plotting

This tutorial presents methods for making plots of data. MATLAB commands introduced include: figure, plot, clf, hold, subplot, legend, xlabel, ylabel, and title. The deliverable is a single page document with different types of MATLAB figures. The PDF file can be downloaded from

http://www.cse.sc.edu/~gatzke/211/matlab4.pdf

6.3.5   MATLAB Tutorials

```
% Tutorial Part 1
% Comments vary from language to language.
% Use comments to explain what your code
% is doing. In MATLAB, anything to the
% right of a % is treated as a comment!
% In the MATLAB editor, everything that
% is a comment is in green

% First, you have to make sure MATLAB
% is in the correct directory. To print
% the current working directory, use pwd
pwd
```

```
% To change the current directory, you
% can use the cd command or the MATLAB
% graphical user interface above the
% script area. It looks like a folder.

% DATA STRUCTURES
% We use variables to represent data
% of different types. Traditional
% data structures include:
% Integers
k=1
p=5

% Real numbers (double precision)
pi=3.141
epsilon=0.001

% and Boolean TRUE FALSE expressions.
% Note that in MATLAB, boolean is
% expressed as 0 or 1 (FALSE adn TRUE)
flag1=(1<0) % This is FALSE, so evaluates to 0
flag2=(1>0) % This is TRUE, so evaluates to 1

% Also note that there are various
% Boolean operators you can use.
% These include <, >, <=, >=, ==, ~=.
% The ampersand & means logical AND
% The pipe | means logical OR
% The double equals == means is equal to
% ~ means complement and xor is XOR
% See: help relop
help relop

% Special data structures
% Arrays - Arrays contain multiple
% pieces of data indexed along one
% or more dimensions for example, a
% vector can be seen as a 1D array
% of real numbers and a matrix can be
% described as a 2D array of real
% numbers. A 1D array is often
% called a vector

% You can use spaces to separate elements
b=[1 2 3 4 5 6] % This makes a row vector with six elements
b(1) % You can access a single element of an array

% You can access multiple elements at once.
b(1:3) % You can access a few elements of an array
 b(4:end) % The end operator is the index for the end of the
array
```

```
% You can change one or more elements.
b(1)=7 % This sets the first element of b to 7
b(2:3)=[6 5] % This sets elements 2 and 3 to 6 and 5

% Semicolons make new rows. You
% can create a column vector this way
c=[1 ; 2 ; 3 ; 4 ; 5 ; 6]

% The length command tells you
% how long a 1D vector is
blen=length(b) % Get the length of b
clen=length(c) % Get the length of c

% The command whos tells you what
% variables are in the MATLAB workspace
% It also shows you the size of the
% variable in memory
whos

% 2D arrays are called matrices.
% They have two indices for row and column.
A=[1 2 3 ; 4 5 6]
A(2,3) % You can access a single element with row
A(:,1) % You can get all the rows from column 1 using :
A(2,:) % You can get all the columns from row 2 using :
A(1:2,2:3) % This gets rows 2 to 3 from column 1 to column 2

% The size command tells you the size
% of an array in the specified dimension.
% Row is first, column is second.
size(A,1)
size(A,2)

% The size command can also return
% two values at once using brackets.
[rows,cols]=size(A)

% Note that you can use N dimensional arrays!
% Some operations like matrix multiplication won't work:
DD(2,2,2,2)=5

% A 3D array is like a stack of matrices.
% A 4D array is a group of 3Ds...

% To forget the values of a variable,
% use the clear command
clear DD
clear D*
% The clear command without a variable
% name clears all the memory.
```

```
% Strings are really just a 1D array
% of single characters.
name='bubba'
city='columbia'
name(2:4)

% You can use arrays of strings
names={'Bobbie','Sue','Thom'}
names(2)

% Most languages handle strings
% and structures differently, so
% watch out when using other programs.
% In some cases, you need to create
% a large matrix or array. The ones and
% zeros commands can help!
rows=200
cols=300
G=ones(rows,cols)
h=zeros(rows,1)

% Note that MATLAB displays all the
% output contents. To suppress this
% output and speed up processing, use
% a semicolon at the end of the line.

G=ones(rows,cols);
h=zeros(rows,1);

% The diary command can copy everything
% done in the editor to a text file

!erase lab1diary.txt % This erases any old file
diary lab1diary.txt % This turns on the diary
% Change the next line to your information!
str='Lab 1 - George Burdell, Sect. 1'
whos % This lists all variables in memory
diary off % This stops writing to the diary

%%
%%%%% Tutorial Part 2 %%%%%
%%%%% Flow Control    %%%%%
%%

clear % Clear the workspace to start

% Flow Control (IF, FOR, WHILE)
% Assignment statements
% Most of you code will be assignment statements.
% When you write a line of code, the name left
% of the = takes on the values of whatever is
% right of the = sign.
```

```
c=2+3 % Assign variable c to be 2+5
c=c+2+c*c % Replace variable c with c+2+c*c

% Some math functions are built in.
d=sqrt(5) % Square root
e=sin(2) % sin function
f=exp(3) % e to the power
g=5^3 % five raised to third power

% Multiple expressions can be evaluated
% at once, be careful of perenthesis.
h=3*(sin(exp(sqrt(6))))^2)

% Order of operations - when making an
% assignment for a complex expression,
% you follow the standard order of
% operations (PEMDAS):
% Please Excuse My Dear Aunt Sally
% Parens, inner first
% Exponents, Powers or root
% Multiply or
% Divide (left to right)
% Add or
% Subtract (left to right)

c= 100 - 10*(2 + 3) + 4
d= 36 / 4*(5 * 3 - 2) + 6
d=((36/4)*((5*3)-2))+6 % Same, but more ()

% If you have doubts, use more
% parens to specify the desired order.
d=((36/4)*((5*3)-2))+6 % Same, but more ()

% FLOW CONTROL
% IF Statements
%
% IF statements allow for sections of
% code to be executed only if a condition
% is met. The condition must evaluate to
% a TRUE or FALSE value.
x=3
if (x>0)    % Check this condition
    x=x^2   % Do this if condition is TRUE
end

% Using the & (AND) operator and the
% | (OR) operator, more complex conditions
% can be specified.

x=-3;
y=-2;
```

```matlab
disp([ x y]) % disp displays the values
if ( (x<0) & (y<0) )
    x=x^2; % Multiple things can be
    y=y^2; % done inside an IF statement.
end
disp([ x y]) % disp displays the values

% Note the indentation of code inside
% the IF statement. This really helps
% code be legible. MATLAB automatically
% tries to indent. You can highlight
% code sections and hit ctrl-i
% IF and ELSE
% Usually, IF syntax includes an ELSE
% condition. Whenever the condition is
% not met, the second section of code
% executes.

x=0;
if (x>0)
    disp('x is strictly positive')
else
    disp('x is 0 or negative')
end

% NOTE - Strings in matlab use the
% single quote next to the ENTER key
% and show up as purple in the editor.
% Usually, IF syntax includes ELSEIF
% conditions. The boolean values are
% checked in order. Whenever a condition
% is met, the corresponding section
% of code is executed. Note that even
% though a second conditional statement
% may evaluate to true, it never gets
% a chance to execute.

x=1;
if (x==0) % The EQUAL to operator is ==
    disp('x is 0')
elseif (x==1)
    disp('x is 1')
elseif (x==2)
    disp('x is 2')
elseif (x==3)
    disp('x is 3')
elseif (x==1)
    disp('x is 1, second time, will not run')
else
    displ('x is not 0, 1, 2, or 3')
end
```

```
% LOOPS
% Loops let you do repetitive stuff easily.
% FOR
% FOR loops are useful when you know
% how many times you may want to run
% the loop before you enter the loop.
% This is especially good for
% manipulating data in an array.
% The following code goes through a
% list of data and replaces each
% value with the value squared.

x=[2 4 3 6 5 7 2 1];
s=length(x);
for k=1:1:s
    x(k)=x(k)^2;
    disp( [ k x(k)] )
end
disp(x)

% In MATLAB, you can start the
% loop counter at any number and
% increment by any value.

x=[]; % This makes an empty variable!
for k=2:5:30
    x=[x k] % This sticks k onto x
end

% Note that array indices in MATLAB
% start at 1. For an array (vector)
% of length s you will get an error
% if you try to access elements 0
% or if you try to access element s+1.

s=length(x);

% The following lines will not work:
% x(0);
% x(s+1);

% WHILE

% WHILE loops continue to evaluate
% until the a boolean value is no
% longer TRUE. These are usually
% used when the number of iterations
% in the loop are not known before
% entering the loop.
```

```
value=rand(); % Random number 0 to 1
while (value>0.001)
    disp(['New value is ' num2str(value) ]);
    value=value*value;
end

% The previous code pics a uniformly
% distributed random number between
% 0 and 1 and repeatedly squares it
% until it is less than 0.001

% NESTED STATEMENTS
% You can have an IF statement
% inside an IF statement:

x=-1;
y=3;
if (x<0)
    if (y<0) % This only executes if x<0
        disp(' x and y are negative');
    else
        disp(' x negative, y positive or 0');
    end
else % The case where x is >=0
    disp(' x is positive or 0');
end

% Note the indentation increases as
% more statements are nested.
% Nested FOR / WHILE

A=[];
for row=1:3
    for col=1:3
        disp([row col]);
        A(row,col)=row+col;
    end % end for columns loop
end % end for rows loop
disp(A)

% Use the following code to make a diary file
% for this lab tutorial section. This gives
% an approximate snapshot to see if you completed
% the lab.

!erase lab2diary.txt % This erases any old file
diary lab2diary.txt % This turns on the diary
% Change the next line to your information!
str='Lab 2 - George Burdell, Sect. 2'
disp([d e f g h ])
disp([row s value x y])
```

```
whos % This lists all variables in memory
diary off % This stops writing to the diary

% Load lab2diary.txt into the editor and print
% out the single page to show you completed the
% lab assignment.

% Lab 2 Bonus:
% Make a script file that does the following:
% 1. Prints your name and section
% 2. Calculates the sum 1^2+2^2+3^2+...500^2
% 3. Calculates the sum 1*2+2*3+3*4+...299*300
% 4. Determines the number of terms required
% in the sum 1+2+3+...to exceed one million
% 5. Makes a matrix A that has 0s on the diagonals,
% 1, above and below, 2, above that, etc
% for a 10x10 matrix. The 4x4 example is:
% A=[0 1 2 3
%    1 0 2 2
%    2 1 0 1
%    3 2 1 0]

%%%%%% Tutorial Part 3    %%%%%%
%%%%%% Linear Algebra and %%%%%%
%%%%%% Root Finding       %%%%%%
%%

clear % Clear the workspace to start

% Variable trace / debugging
% When you have an error, the
% error may be apparent or hidden.
% An apparent error may cause the
% program to stop and report the
% offending line. The error may
% be a syntax error or a coding
% logic error (array index problem,
% divide by zero, etc). Hidden
% errors cause the code to run in
% unintended ways.

A=ones(2,3);
AA=zeros(4,5);

% Either way, you may need to
% "trace" variable values to make
% sure the program is performing
% as expected. In MATLAB, you can
% print variable values just by
% using the name of the variable
% without a semicolon.
```

A

```
% or use whos to get information
% about a variables, like the size.

whos
whos A*

% User defined scripts
% In MATLAB, you can save a string
% of commands in a textfile with a
% .m extension on the end of the file
% name. Typing the name of the file
% at the command prompt will cause
% the commands to be executed as if you
% were typing commands at the prompt.
% MATLAB has special operations for
% vectors. You can easily add vectors.

b=[1 2 3]; % Define a vector
c=[4 5 6]; % Define a vector
d=b+c       % You can add vectors
e=2*b+3*c   % You can multiply by a scalar

% MATLAB can do dot and cross products

dot(b,c)    % Dot product of vectors
cross(b,c) % Cross product of vectors

% MATLAB can also do matrix operations

A=[1 2 3 ; 4 5 6 ; 7 8 9]; % Define matrix
det(A) % Determinant of a matrix
AT=A'   % Transpose of a matrix, swaps rows
        % and columns.

% MATLAB can find the matrix inverse
inv(A) % Inverse of a matrix
A*inv(A) % Should be identity matrix

% The eye command creates an identity matrix

eye(4) % Identity matrix size 4

% In many cases, you want to solve sets
% of linear equations in the form
% A*x=b
% Define a nonsingular matrix

A=[-1 2 3 ; 4 -5 6 ; 7 8 -9];
det(A) % If this is 0, A is singular
```

```
x=[1 ; -2 ; 4]; % Known x value
b=A*x;    % This sets b if x is known
inv(A)*b % This should be x

% Solve the following problem:
% 1*x + 2*y = 3
% 4*x - 5*y = 6
%
% In form Ax=b
% [ 1 2 ] [x] = -10
% [ 4 -5 ] [y] = 77
% The solution vector x contains x and y
% or sometimes called x1 and x2

A=[1 2 ; 4 -5]
b=[-10 ; 77]
x=inv(A)*b

% MATLAB also does matrix multiplication
% if the matrices are conformable.

B=[2 3 4 ; 5 6 7];
C=A*B;

% B*A is not conformable because the
% inner dimensions do not match.
% Matrix A is 2x2 and B is 2x3
% You cannot multiply 2x3 * 2x2 but
% you can multiply 2x2 * 2x3!
% You can define functions inline

f=inline('x*x-3');

f(2) % Evaluates the function at x=2

% The fzero function will find a root
% of an equation where f(x)=0

fzero(f,1) % fzero finds where f(x)=0

% The polyval function evaluates a
% polynomial in the form
% p_n x^n + p_n-1 x^n-1 +...+ p_1 x + p_0
% Polynomial for 1 x^2 + 2 x - 1

p=[1 2 -1];
polyval(p,0) % Evaluate at x=0
polyval(p,1) % Evaluate at x=1

% Polynomial for 1 x^3 + x^2 - 2 x + 0
p2=[ 1 1 -2 0]
```

```
% The roots command finds the solutions
% for polynomials where p(x)=0
roots(p)
roots(p2)

% For the polynomial 1x^2+2x+3
roots([1 2 3]) % Imaginary solutions!

% You can quickly make a vectors starting
% at -3 and going to 2 with 0.2 jumps as:
x=[-3:.1:2]';

% Make a matrix Y where:
% column 1 has polynomial 1 values
% column 2 has polynomial 2 values
% column 3 has all zeros
Y=[polyval(p,x) polyval(p2,x) x*0]

% Make a plot to hand in:
clf % clear the current figure
plot(x,Y) % Plots y and x

% Change the following line to your information
title('Lab 3 by George P. Burdell, Section 0')

% Put the contents of some values on the plot
data=[[c d e ] ; [A B C zeros(2,1)] ]

% Write data on plot to show you completed the lab
text(-2,-10,num2str(data))

% Note that num2str can make a string from a number.
% Print out the figure to hand in

% Lab 3 Bonus:
% Make a matlab script that does the following:
% 1. Prints your name and section
% 2. Make a 40x40 matrix with random numbers
%     from the normal distribution.
% 3. Using that matrix, calculate the total
%     of the positive elements and the total
%     of the negative elements.
% 4. Using that matrix, calculate the number
%     of positive and negative elements.
% 5. Using that matrix, determine the minimum
%     and maximum values
% 6. Takes the matrix and replaces any values
%     less than -2 or greater than 2 with 0.
%%
%%%%% Tutorial Part 4 %%%%%
%%%%% Figures %%%%%
```

```
%%
% In this section, you will make multiple
% figures. Rather than print each out
% individually, you are encouraged to
% put them all in a single page Word
% document.
clear % Clear the workspace to start

% C(t) = a*exp(-k*t)
% ln(C(t)) = ln(a) + (-k) * t
% y = b + m x
%
% Make some data with noise
T=[0:1:10]';
C=120*exp(-0.34*T)+5*randn(11,1);
C=C.*(C>0)+.001;

% Plot the data using plot
% plot(xvals,yvals,linetype)
clf % This clears the plot
plot(T,C,'x') % This plots x and y data

% Note, the plotting line type can be changed
% with various values. Instead of 'x' as
% the type, 'r*' would use red stars.

% Add a title and labels
ylabel('Concentration (mol/L)')
xlabel('Time (hr)')
title('Concentration vs. Time')
% Keep a copy of this plot to turn in.

% You can copy paste the plot into a Word
% document. From the figure window, select
% Edit -> Copy figure then paste into Word.
% For noisy data, you need a best fit line.
% Regress the data using regress(yvals,xvals)
% y = b + m x
% Need to add a column of ones to get the b
% coeficient values.
[vals,conf]=regress(log(C),[ones(11,1) T])
a=exp(vals(1))
k=-vals(2)

% Are these values close to 120 and 0.34?
% Use the found values to find model values
% ln(C(t)) = ln(a) + (-k) * t
% y = b + m x
clf;
ymodel=log(a)-k*T;
ytrue=log(120)-0.34*T;
plot(T,log(C),'x',T,ymodel,T,ytrue,'--')
```

```matlab
% Add a legend
legend('Data','Model','True Model')

% Add a title and labels
ylabel('ln(Conc. (mol/L))')
xlabel('Time (hr)')
title('Concentration vs. Time')

% Keep a copy of this plot to turn in.
% Now, use the model values for a and k
% to make a plot against the original
% data before the transformation.
clf
Cmodel=a*exp(-k*T);
Ctrue=120*exp(-.34*T);
plot(T,C,'x',T,Cmodel,T,Ctrue,'--')
legend('Data','Model','True Model')

% Add a title and labels
ylabel('Concentration (mol/L)')
xlabel('Time (hr)')
t=title('Concentration vs. Time')

% You can script changes to text
% variable t was set to be a text
% handle which identifies which
% part of the figure you set to
% be the title. You can change
% various values of the text
set(t,'FontSize',14,'Color',[1 0 0])

% Use get to see what values are
% possible to change
get(t)

% Keep a copy of this plot to turn in.
% In some cases, you may have multiple
% experiments (replicates) conducted
% under different experimental conditions.
% You can use error bars to show the
% distribution of the data.
% Make three sets of noisy data at three
% different values of x using a FOR loop.

x=[1 2 3];
for k=1:3
data1(:,k)=10+2.0*x(k)+1*randn(10,1)
data2(:,k)=22+1.8*x(k)+2*randn(10,1)
data3(:,k)=34+1.5*x(k)+3*randn(10,1)
end
```

```
% Get the mean and standard deviation into
% two matrices for plotting.

Y=[mean(data1)' mean(data2)' mean(data3)'];
dev=[std(data1)' std(data2)' std(data3)'];

% MATLAB allows lists of differently sized
% items using curly braces.
linetypes={'rx','bo','*'}

% Use the hold command to allow plotting
% over an existing plot. The command
% hold off allows the next plot command
% to wipe out what is there. Using
% hold on allows the next plot to overwrite
% any existing plot.

clf
hold off
for k=1:3
    errorbar(x,Y(:,k),dev(:,k),dev(:,k),linetypes{k})
    hold on
end

axis([0 4 -5 45])
legend('\mu_1=10','\mu_2=22','\mu_3=34', ...
'Location','SouthEast')

ylabel('\sigma^2 values')

% Note that subscript and superscript
% as well as greek letters are possible

hold off
title('Three Experiments')

% Keep a copy of this plot to turn in.

% You can make a histogram of random data
% using the hist command. The second
% input argument is where the bins are centered.
% The bar command makes similar bar graphs.

clf
testgrades=50+13*randn(100,1);
hist(testgrades,[2.5:5:97.5])
tavg=mean(testgrades)
str=['Test grades, \mu=' num2str(tavg) ]
title(str)
% Keep a copy of this plot to turn in.

% You can make 3D plots using mesh or plot3
x=[-10:.5:10];
```

```
y=[-10:1:10];
Z=[];
for row=1:length(x)
      for col=1:length(y)
            Z(row,col)=x(row)*y(col);
      end
end

clf % Clear the figure
mesh(x,y,Z') % Note the Z transpose!
xlabel('x')
ylabel('y')
zlabel('z')
title('z=x*y')

% Keep a copy of this plot to turn in.
% Again, make some data for three
% different experiments. This time
% the data is dynamic for a simple
% system. The input to the system is
% u wchich changes from 0 to 1 at time
% t=0.

t=[-1:.01:10]'; % Make a time vector
u=t>=0; % Make the input u vector

% Make the three sets of data for the
% response of the dynamic system

y1=2.*u-2*exp(-t/1).*u;
y2=2.*u-2*exp(-t/2).*u;
y3=2.*u-2*exp(-t/3).*u;
subplot(2,1,1)
plot(t,u)
axis([-1 10 -.1 1.1])
ylabel('Input value u(t)')
title('Plot by George P. Burdell')
subplot(2,1,2)
plot(t,y1,t,y2,'--',t,y3,'-.')
axis([-1 10 -.1 2.1])
title('\tau dy/dt + y = 2 u(t)')
ylabel('Response y(t)')
xlabel('Time (s)')
legend('\tau=1','\tau=2','\tau=3')
%Print out the plot to turn in.

% User defined function
% If you continually are doing
% the same procedure, you can
% generalize the procedure to
% make your own function. Some
```

```
% functions are built in (sin,
% exp, length, etc).
% The function takes input
% arguments, performs some
% operations, and returns output
% values. In MATLAB, you put
% your function in a text file
% with a .m extension.
% Put the following four lines
% in a file and save it as stat.m

%function mean = stat(x)
%STAT Interesting statistics.
%n = length(x);
%mean = sum(x) / n;

% You could call the funciton like
% any built-in function:
% stat([ 1 2 3 4 3 4 5])

% You can have more interesting
% functions that return
% multiple outputs at once.

% Replace your stat.m funciton file
% with the following five lines:

%function [mean,stdev] = stat(x)
% % STAT Interesting statistics.
%n = length(x); % How much input?
%mean = sum(x) / n; % Get the average
%stdev = sqrt(sum((x - mean).^2)/n);

% Scope of variables
% Variable scope is important!
% Inside functions, you may use
% new variables. These are often
% called local variables. In the
% previous example, n is a local
% variable. n takes a value when
% the function is called. If
% variable n had a value outside
% of the function, it would not
% be changed when stat is called.

n=5
% stat([1 2 3 6 7 8])
n

% Variables can be defined as
% global. This means that they
```

```
% can be changed inside a
% subroutine, assuming the
% subroutine knows it is a
% global variable.

clear n
global n
n=5
%stat([1 2 3 4 5 5 6 7 8])
n

% A function can call itself. This
% is called a rcursive function.

% function out=fact(x)
% if (isreal(x))
% if (x==1)
%     out=1
% else
%     out=x*fact(x-1)
% end
% end

% Everything to this point has been with respect to
% procedural programming. Computer Scinece classes
% will often discuss object-oriented programming.
% This is a methodology that considers all data
% as objects. These objects all have a class. One
% object may be a subclass of another object. For
% example, you may have a class student. There may
% be a subclass undergraduate and a subclass
% graduate. All students should have a name, but
% undergraduates would have class standing and
% graduate students would have advisors. Procedures
% can be written for each class. MATLAB is not easy
% to use for object oriented programming. Other
% languages such as Java may be a better choice
% for object oriented programming.
```

Chapter 7
Linear Algebra

7.1 Introduction

Linear algebra is a very powerful tool. When students take a mathematics course in linear algebra, they often fail to realize how useful the methods can be for real-world applications in engineering and science. If a problem can be expressed as a set of linear algebraic equations, rapid solution is available for problems with thousands of variables. Linear algebra nomenclature also allows large complex problems to be expressed in simple, elegant terms. Additionally, linear algebra plays a key role in solution of a wide range of problems, including parametric regression, solution of multivariate nonlinear equations, and solution of differential equations. To begin, let us start with a simple example problem.

Example 7.1 Linear Algebra Example System

Two distillation columns in series with an additional feed stream mixing in with the bottoms stream of the first column. The flow rate of three streams is unknown. As indicated in Fig. 7.1, the flow rate of streams x, y, and z is unknown. No reaction is taking place. The steady-state flow rates must be calculated.

The general form basic mass balance is given as:

$$accumulation = in - out + created - destroyed$$

One may write a mass balance on the first column. In this case, one may assume steady-state, meaning that $accumulation = 0$.

$$0 = 100 - 40 - x$$

A mass balance on the mixing point between the two columns is written as:

$$0 = x + 30 - y$$

© Springer Nature Switzerland AG 2022
E. Gatzke, *Introduction to Modeling and Numerical Methods for Biomedical and Chemical Engineers*, https://doi.org/10.1007/978-3-030-76449-4_7

Fig. 7.1 Two distillation
columns in series

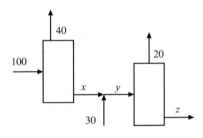

Flowrates in *kmol / hr*

A mass balance on the second column is given by:

$$0 = y - 20 - z$$

Three resulting linear equations are:

$$0 = 100 - 40 - x$$
$$0 = x + 30 - y$$
$$0 = y - 20 - z$$

Note That You Could Write Too Many Equations You could write an overall balance around the entire system:

$$0 = 100 - 40 - 20 - z$$

With four equations and three variables, this would be called an over specified system of equations. We will stick with the first three equations from above for now.

Note That These Are Linear Equations The unknown variables have constant linear coefficients, and nonlinear terms do not appear (no x^2, no $\sqrt{x}$, no e^x).

You can rearrange the set of three equations (excluding the overall balance equation) to get all the variable terms on the left side and the constants on the right. After some algebraic manipulation, the set of equations can be written as:

$$
\begin{aligned}
1x + 0y + 0z &= 60 \\
-1x + 1y + 0z &= 30 \\
0x - 1y + 1z &= -20
\end{aligned}
\tag{7.1}
$$

As we will see later, this can be more compactly written as:

$$\underline{\underline{A}}\,\underline{x} = \underline{b}$$

You may already realize that the solution to this problem is $x = 60$, $y = 90$, and $z = 70$. For more complex systems, this is not quite so easy. To solve the three linear equations simultaneously in a general manner, you can perform row reduction using three possible row operations.

7.2   Solution by Row Reduction

There are three rules that can be used to rapidly find a solution to a set of linear equations. These rules define how one manipulates a set of equations. Given a set of equations, the following steps can be taken:

1. Add (or subtract) one row to (or from) another.
2. Multiply or divide a row by a scalar value (any real scalar $\neq 0$).
3. Swap position of rows.

Typically, you would perform these operations until you have a triangular representation (all zeros above or below the diagonal). The triangular form allows for quick determination of the actual numeric solution.

Example 7.2 (Row Reduction) The set of linear equations in Eq. 7.1 can be compactly written using only the coefficients as:

$$
\begin{array}{ccc|c}
1 & 0 & 0 & 60 \\
-1 & 1 & 0 & 30 \\
0 & -1 & 1 & -20
\end{array}
$$

We need to perform steps 1–3 to get the system of equations in triangular form with ones on the diagonal and zeros below the diagonal, like:

$$
\begin{array}{ccc|c}
1 & a & b & d \\
0 & 1 & c & e \\
0 & 0 & 1 & f
\end{array}
$$

We can look at the original system of equations and realize that we must get zeros in position 2,1 (row 2, column 1) and position 3,2 (row 3, column 2). You can multiply row 2 by -1 using Rule 2:

$$
\begin{array}{ccc|c}
1 & 0 & 0 & 60 \\
1 & -1 & 0 & -30 \\
0 & -1 & 1 & -20
\end{array}
$$

Next, swap position of rows 2 and 3 using Rule 3 to get:

$$\left[\begin{array}{ccc|c} 1 & 0 & 0 & 60 \\ 0 & -1 & 1 & -20 \\ 1 & -1 & 0 & -30 \end{array}\right]$$

Then, subtract row 1 from row 3 using Rule 1 to get:

$$\left[\begin{array}{ccc|c} 1 & 0 & 0 & 60 \\ 0 & -1 & 1 & -20 \\ 0 & -1 & 0 & -90 \end{array}\right]$$

Then, multiply rows 2 and 3 by -1 using Rule 2:

$$\left[\begin{array}{ccc|c} 1 & 0 & 0 & 60 \\ 0 & 1 & -1 & 20 \\ 0 & 1 & 0 & 90 \end{array}\right]$$

Subtract row 2 from row 3 using Rule 1 again to get:

$$\left[\begin{array}{ccc|c} 1 & 0 & 0 & 60 \\ 0 & 1 & -1 & 20 \\ 0 & 0 & 1 & 70 \end{array}\right]$$

Now, all coefficients below the diagonal are 0. The solution can be found quickly because of the diagonal structure. Remember that the numerical values here represent three equations that have been transformed to:

$$1x + 0y + 0z = 60$$
$$0x + 1y - 1z = 20$$
$$0x + 0y + 1z = 70 \qquad (7.2)$$

From equation 3 (row 3), $z = 70$. Since z is now known, one may use equation 2 (row 2) $y - z = 20$ to find that $y = 90$. Equation 1 (row 1) gives $x = 60$, so the overall solution is $x = 60$, $y = 90$, and $z = 70$.

CHECK SOLUTIONS You can plug your solution back into the original three equations and verify that the equations are satisfied. **THIS WILL HELP YOU ON EXAMS**. If you assign numerical values to the variables, the equations should be satisfied.

The general Gaussian elimination or row reduction method specifies that you start with column 1 and perform operations until a 1 is on the diagonal and all coefficients below the diagonal are 0. The method specifies that you then move to column 2 and perform operations until a 1 is on the diagonal and all coefficients below the diagonal are zero. This repeats until the diagonal elements are all 1.

7.3   Linear Equations: Special Cases

In general, there are three possibilities for a "square" set of linear equations.

7.3.1   Case A: One Solution

Consider a simpler system: $x + y = 1$ and $x - y = 1$. Graphically, you can plot the two lines and look for the intersection of two lines which occurs at $x = 1$, $y = 0$. The system of equations is:

$$x + y = 1$$
$$x - y = 1$$

These two equations involving two variables can also be represented as:

$$\begin{array}{cc|c} 1 & 1 & 1 \\ 1 & -1 & 1 \end{array}$$

Subtracting row 1 from row 2 gives:

$$\begin{array}{cc|c} 1 & 1 & 1 \\ 0 & -2 & 0 \end{array}$$

This implies $-2y = 0$ or $y = 0$ and $x + y = 1$ or $x = 1$ as you already realized.

In three dimensions (3 unknowns), each row represents a plane in a three-dimensional space. Two equations in 3D can intersect to give a line. A line in 3D can intersect with a third plane to give a point, the single solution (in a single solution case).

7.3.2   Case B: No Solution

Consider the system $x + y = 1$ and $x + y = 2$. Graphically, this represents two lines that never intersect.

$$\begin{array}{cc|c} 1 & 1 & 1 \\ 1 & 1 & 2 \end{array}$$

Note that column 1 and column 2 are identical. Subtracting row 1 from row 2 gives:

$$\begin{array}{cc|c} 1 & 1 & 1 \\ 0 & 0 & 1 \end{array}$$

You know that $0x + 0y = 1$ cannot be true. For a "square" system, if Gaussian elimination results in a 0 on the diagonal, this may be the case.

7.3.3   Case C: Many Solutions

Consider the system $x + y = 1$ and $2x + 2y = 2$. Graphically, this represents two lines that are coincident.

$$\begin{array}{cc|c} 1 & 1 & 1 \\ 2 & 2 & 2 \end{array}$$

Subtracting twice the value of row 1 from row 2 gives:

$$\begin{array}{cc|c} 1 & 1 & 1 \\ 0 & 0 & 0 \end{array}$$

These equations are consistent; $0x + 0y = 0$ and $x + y = 1$ are consistent. There is no single solution, as many solutions make the equation $x + y + 1$ consistent.

7.3.4   Non-Square Systems

The original example was for a "square" system with 3 unknowns and 3 equations. You may often end up with more (or fewer) equations than unknowns.

Consider the original set of equations:

$$1x + 0y + 0z = 60$$
$$-1x + 1y + 0z = 30$$
$$0x - 1y + 1z = -20$$

One additional equation can be specified using a mass balance on the entire system, $0 = 100 + 30 - 40 - 20 - z$.

$$1x + 0y + 0z = 60$$
$$-1x + 1y + 0z = 30$$

$$0x - 1y + 1z = -20$$
$$0x + 0y + 1z = 70 \tag{7.3}$$

These four linear equations are not "linearly independent." You can test this by using row operations to make two rows identical. Simultaneously adding row 1 and row 3 to row 2 will make row 2 the same as row 4.

$$1x + 0y + 0z = 60$$
$$0x + 0y + 1z = 70$$
$$0x - 1y + 1z = -20$$
$$0x + 0y + 1z = 70 \tag{7.4}$$

This set of equations can still be satisfied using the original solution $x = 60$, $y = 90$, and $z = 70$. In other cases, having more equations than unknowns will complicate the solution process a bit. However, the solution of the over-specified system is quite useful for many engineering problems.

7.4   Vectors

A group of unknown (or known) values can be "stacked" to form a vector. In the example problem, the unknowns x, y, and z can be described by the vector $\underline{x}$:

$$\underline{x} = \begin{bmatrix} x \\ y \\ z \end{bmatrix}$$

The solution to the problem has a known value and can be written as a vector of constant values:

$$\underline{x} = \begin{bmatrix} 60 \\ 90 \\ 70 \end{bmatrix}$$

Note that the underbar can be used to distinguish between $\underline{x}$ (the vector) and x the single unknown value. Typically, in linear algebra, a vector stacks items vertically and is more formally called a column vector.

Column Vector—A column vector in linear algebra is a one-dimensional construct of either variables or constant values.

Realize that a vector is NOT limited to 2 or 3 unknowns. For example, the concentration of a drug in dermal tissue measured every 1mm of depth can be seen as a vector:

$$\underline{c} = \begin{bmatrix} 0.11 \\ 0.13 \\ 0.16 \\ 0.19 \\ 0.22 \end{bmatrix}$$

Similarly, a complex problem may involve tens or hundreds of unknowns. The temperature of a substance in a pipe could be modeled along a length l from $l = 0$ to $l = L$ as:

$$\underline{T} = \begin{bmatrix} T_{l=0} \\ \vdots \\ T_{l=L} \end{bmatrix}$$

When considering a vector in linear algebra, in most cases, the vector is a column vector with vertically stacked elements.

7.4.1   Dot Product

The dot product is an important value calculated from two vectors of the same number of elements.

Dot Product—The dot product of two vectors is the sum of the product of the individual elements.

$$\underline{x} \cdot \underline{y} = \sum_i x_i \, y_i$$

The dot product of two vectors can only be calculated if both vectors have the same number of elements. Typically, it is assumed that both vectors are column vectors. If two vectors represent directions at right angles to one another, the dot product of the two vectors will be 0. Consider the vectors $\begin{bmatrix} 1 \\ 0 \end{bmatrix}$ and $\begin{bmatrix} 0 \\ 1 \end{bmatrix}$:

$$\begin{bmatrix} 1 \\ 0 \end{bmatrix} \cdot \begin{bmatrix} 0 \\ 1 \end{bmatrix} = (1) \times (0) + (0) \times (1) = 0$$

Example 7.3 (Dot Product) Here are a few examples of dot product calculation:

$$\underline{x} \cdot \underline{x} = \begin{bmatrix} x \\ y \\ z \end{bmatrix} \cdot \begin{bmatrix} x \\ y \\ z \end{bmatrix} = x^2 + y^2 + z^2$$

$$\begin{bmatrix} 1 \\ 2 \\ 3 \end{bmatrix} \cdot \begin{bmatrix} x \\ y \\ z \end{bmatrix} = 1x + 2y + 3z$$

$$\begin{bmatrix} 1 \\ 2 \\ 3 \end{bmatrix} \cdot \begin{bmatrix} 4 \\ 5 \\ 6 \end{bmatrix} = 1 \times 4 + 2 \times 5 + 3 \times 6 = 32$$

7.5   The Matrix

A matrix can be seen as similar to a vector but having two dimensions. One way to think of a matrix is as a group of vectors augmented together. A matrix has a size, $m \times n$ representing m rows and n columns. The values for m and n are sometimes written as subscripts for the matrix. For example, the 2x3 matrix $\underline{\underline{A}}_{2 \times 3}$ with two rows and three columns may have values as follows:

$$\underline{\underline{A}}_{2 \times 3} = \begin{bmatrix} a_{1,1} & a_{1,2} & a_{1,3} \\ a_{2,1} & a_{2,2} & a_{2,3} \end{bmatrix}$$

Each of the six elements has two indices. The first index is the row, and the second is the column. For the applications in this book, a matrix will have constant coefficient values. Some example matrices are:

$$\underline{\underline{A}}_{2 \times 3} = \begin{bmatrix} 0 & -2 & 1 \\ 5 & 1 & 0.2 \end{bmatrix} \quad \underline{\underline{B}}_{3 \times 3} = \begin{bmatrix} 6 & 0 & 0 \\ -2 & 0 & -1 \\ 3 & -1 & 5 \end{bmatrix}$$

Matrix—A matrix is a two-dimensional construct where each element may be indexed by both row and column values.

Square Matrix—A square matrix is a matrix with equal indices ($m = n$). This means that the number of rows and the number of columns are the same.

Note: a vector can be seen as a special matrix having only 1 column.

Identity Matrix—The identity matrix has values of ones on the diagonal and zeros elsewhere. It is defined as $\underline{\underline{I}}$, and for a square matrix, $\underline{\underline{A}}\,\underline{\underline{I}} = \underline{\underline{A}}$ and $\underline{\underline{I}}\,\underline{\underline{A}} = \underline{\underline{A}}$. For a three-by-three system, the identity matrix is given by:

$$\underline{\underline{I}} = \begin{bmatrix} 1 & 0 & 0 \\ 0 & 1 & 0 \\ 0 & 0 & 1 \end{bmatrix}$$

Matrix Transpose—The matrix transpose operator swaps the indices of a matrix (or a vector).

Consider the general matrix $\underline{\underline{A}}_{2\times3}$ as before. The matrix transpose of $\underline{\underline{A}}$ is given in general as:

$$\left(\underline{\underline{A}}_{2\times3}\right)^T = \begin{bmatrix} a_{1,1} & a_{1,2} & a_{1,3} \\ a_{2,1} & a_{2,2} & a_{2,3} \end{bmatrix}^T = \begin{bmatrix} a_{1,1} & a_{2,1} \\ a_{1,2} & a_{2,2} \\ a_{1,3} & a_{2,3} \end{bmatrix}$$

Example 7.4 (Matrix Transpose) For the 2×2 matrix $\underline{\underline{A}}$,

$$\underline{\underline{A}} = \begin{bmatrix} 1 & 2 \\ 3 & 4 \end{bmatrix}$$

$$\underline{\underline{A}}^T = \begin{bmatrix} 1 & 3 \\ 2 & 4 \end{bmatrix}$$

Finally, one can take the transpose of a vector. For the vector $\underline{x} = \begin{bmatrix} x \\ y \\ z \end{bmatrix}$,

$$\underline{x}^T = [x\ y\ z] = \begin{bmatrix} x \\ y \\ z \end{bmatrix}^T$$

Row Vector—The transpose of a vector is also known as a row vector.

7.5.1 Matrix Multiplication

Two matrices can be multiplied together; for example, $\underline{\underline{A}}_{m \times n}$ can be multiplied by $\underline{\underline{B}}_{n \times j}$. Matrix $\underline{\underline{A}}$ has m rows and n columns, while $\underline{\underline{B}}$ has n rows and j columns.

$$
\underline{\underline{A}}_{m \times n} = \begin{bmatrix} \cdots & r_1 & \cdots \\ \cdots & r_2 & \cdots \\ & \vdots & \\ \cdots & r_m & \cdots \end{bmatrix}
$$

Here, each row in matrix $\underline{\underline{A}}$ up to r_m is a row vector with n elements. Alternatively, consider matrix $\underline{\underline{B}}$:

$$
\underline{\underline{B}}_{n \times j} = \begin{bmatrix} \vdots & \vdots & & \vdots \\ c_1 & c_2 & \cdots & c_j \\ \vdots & \vdots & & \vdots \end{bmatrix}
$$

Here, each column up to column c_j is a vector (column vector) with n elements.

Matrix Multiplication—To compute the matrix product $\underline{\underline{A}}_{m \times n} \underline{\underline{B}}_{n \times j}$ or simply just $\underline{\underline{A}}\ \underline{\underline{B}}$,

$$
\underline{\underline{A}}_{m \times n} \underline{\underline{B}}_{n \times j} = \begin{bmatrix} r_1^T \cdot c_1 & r_1^T \cdot c_2 \cdots & r_1^T \cdot c_j \\ r_2^T \cdot c_1 & r_2^T \cdot c_2 \cdots & r_2^T \cdot c_j \\ \vdots & \vdots & \vdots \\ r_m^T \cdot c_1 & r_m^T \cdot c_2 \cdots & r_m^T \cdot c_j \end{bmatrix}
$$

To compute the resulting matrix product of $\underline{\underline{A}}_{m \times n} \underline{\underline{B}}_{n \times j}$, the result will have j columns. The first column of the result is computed by taking the dot product of $\underline{\underline{B}}_{1 \times j}$ (first column of $\underline{\underline{B}}$) with the transpose of all the rows of $\underline{\underline{A}}$. The second column of the result is computed by taking the dot product of $\underline{\underline{B}}_{2 \times j}$ (second column of $\underline{\underline{B}}$) with the transpose of all the rows of $\underline{\underline{A}}$. Repeat up to the jth column of $\underline{\underline{B}}$, which produces the jth column of the result.

Conformable Matrices—Conformable matrices can be multiplied together. In order to multiply $\underset{=m \times n}{A} \underset{=n \times j}{B}$ the "inner"İ dimensions must be equal. For $\underset{=m \times n}{A} \underset{=n \times j}{B}$, matrix A has m columns and matrix B has n rows so the result will have m columns and n rows. The number of columns in the first matrix must equal the number of rows in the second matrix.

Example 7.5 (Matrix Multiplication) Matrix Multiplication Examples:

$$\begin{bmatrix} 1 & 2 \\ 3 & 4 \end{bmatrix}\begin{bmatrix} 5 & 6 \\ 7 & 8 \end{bmatrix} = \begin{bmatrix} 5+14 & 6+16 \\ 15+28 & 18+32 \end{bmatrix} = \begin{bmatrix} 19 & 22 \\ 43 & 50 \end{bmatrix}$$

$$\begin{bmatrix} -1 & 2 \\ 1 & 1 \end{bmatrix}\begin{bmatrix} 4 \\ 5 \end{bmatrix} = \begin{bmatrix} -4+10 \\ 4+5 \end{bmatrix} = \begin{bmatrix} 6 \\ 9 \end{bmatrix}$$

$$\begin{bmatrix} -1 & 2 \\ 1 & 1 \end{bmatrix}\begin{bmatrix} x \\ y \end{bmatrix} = \begin{bmatrix} -x+2y \\ x+y \end{bmatrix}$$

$$\begin{bmatrix} 2 & 3 \\ 1 & -1 \\ 5 & 0 \end{bmatrix}\begin{bmatrix} 2 & 0 \\ -2 & 1 \end{bmatrix} = \begin{bmatrix} 4-6 & 3 \\ 2+2 & -1 \\ 10+0 & 0 \end{bmatrix} = \begin{bmatrix} -2 & 3 \\ 4 & -1 \\ 10 & 0 \end{bmatrix}$$

One final note on matrix multiplication is that matrix multiplication does not commute. This means that in general, $\underline{A}\,\underline{B} \neq \underline{B}\,\underline{A}$. This is especially true for non-square matrices that may be conformable in one order but cannot be multiplied together when the order is switched. However, there are some cases where you can switch the order of the matrices. This is true when multiplying by the identity matrix:

$$\begin{bmatrix} 1 & 2 \\ 3 & 4 \end{bmatrix}\begin{bmatrix} 1 & 0 \\ 0 & 1 \end{bmatrix} = \begin{bmatrix} 1 & 0 \\ 0 & 1 \end{bmatrix}\begin{bmatrix} 1 & 2 \\ 3 & 4 \end{bmatrix} = \begin{bmatrix} 1 & 2 \\ 3 & 4 \end{bmatrix}$$

7.5.2   Determinant of a Matrix

The determinant of a square matrix can be calculated. Assume the following:

- The determinant of a single value is the value itself, $\det(a) = a$.
- The determinant of a 2 × 2 matrix $\det\begin{bmatrix} a & b \\ c & d \end{bmatrix}$ is the value $ad - bc$.
- The determinant of a $N \times N$ matrix can be determined by the sum:

$$\det\left(\underline{\underline{A}}\right) = \sum_{i=1}^{N} (-1)^{i+1} \left(\underline{\underline{A}}_{1,i}\right) \det\left(\underline{\underline{B}}_i\right)$$

where $\left(\underline{\underline{B}}_i\right)$ is the square matrix formed from matrix $\underline{\underline{A}}$ by removing the first row of $\underline{\underline{A}}$ and the ith column of $\underline{\underline{A}}$.

The determinant of a matrix can be illustrated using the following example.

Example 7.6 (Matrix Determinant) A 3×3 matrix determinant can be found as the sum of three 2×2 matrix determinants.

$$\det \begin{bmatrix} 1 & 2 & 3 \\ 4 & 5 & 6 \\ 7 & 8 & 9 \end{bmatrix} = (-1)^2\,(1) \det \begin{bmatrix} 5 & 6 \\ 8 & 9 \end{bmatrix} +$$

$$(-1)^3\,(2) \det \begin{bmatrix} 4 & 6 \\ 7 & 9 \end{bmatrix} +$$

$$(-1)^4\,(3) \det \begin{bmatrix} 4 & 5 \\ 7 & 8 \end{bmatrix}$$

$$= (1)(45 - 48) - (2)(36 - 42) + (3)(32 - 35)$$

$$= -3 - (-12) + (-9)$$

$$= 0$$

7.5.3   Cross Product

Consider vectors defined in three dimensions. We define unit vectors pointing in the x, y, and z directions as:

$$\underline{i} = \begin{bmatrix} 1 \\ 0 \\ 0 \end{bmatrix} \quad \underline{j} = \begin{bmatrix} 0 \\ 1 \\ 0 \end{bmatrix} \quad \underline{k} = \begin{bmatrix} 0 \\ 0 \\ 1 \end{bmatrix}$$

The cross product of two vectors creates a new vector with special properties.

Cross Product—The cross product of two vectors $\underline{a}$ and $\underline{b}$ may be determined by evaluating:

$$\det \begin{bmatrix} \underline{i} & \underline{j} & \underline{k} \\ a_1 & a_2 & a_3 \\ b_1 & b_2 & b_3 \end{bmatrix}$$

The resulting value is a vector that points in a direction orthogonal to the plane defined by $\underline{a}$ and $\underline{b}$.

Example 7.7 (Cross Product) Consider the two vectors $\underline{a}$ and $\underline{b}$ given as:

$$\underline{a} = \begin{bmatrix} 1 \\ 2 \\ 3 \end{bmatrix} \quad \underline{b} = \begin{bmatrix} 4 \\ 5 \\ 6 \end{bmatrix}$$

The cross product $\underline{c}$ is determined by evaluating:

$$\underline{c} = \det \begin{bmatrix} \underline{i} & \underline{j} & \underline{k} \\ 1 & 2 & 3 \\ 4 & 5 & 6 \end{bmatrix}$$

This becomes

$$\underline{i}\,\det \begin{bmatrix} 2 & 3 \\ 5 & 6 \end{bmatrix} - \underline{j}\,\det \begin{bmatrix} 1 & 3 \\ 4 & 6 \end{bmatrix} + \underline{k}\,\det \begin{bmatrix} 1 & 2 \\ 4 & 5 \end{bmatrix}$$

Or more after evaluation of the 2×2 determinants,

$$\underline{i}\,(-3) - \underline{j}\,(-6) + \underline{k}\,(-3)$$

$$(-3) \begin{bmatrix} 1 \\ 0 \\ 0 \end{bmatrix} + (6) \begin{bmatrix} 0 \\ 1 \\ 0 \end{bmatrix} + (-3) \begin{bmatrix} 0 \\ 0 \\ 1 \end{bmatrix} = \begin{bmatrix} -3 \\ 6 \\ -3 \end{bmatrix}$$

You can check that $\underline{c}$ is orthogonal to $\underline{a}$ and $\underline{b}$ using the dot product:

$$\begin{bmatrix} 1 \\ 2 \\ 3 \end{bmatrix} \cdot \begin{bmatrix} -3 \\ 6 \\ -3 \end{bmatrix} = -3 + 12 - 9 = 0$$

$$\begin{bmatrix} 4 \\ 5 \\ 6 \end{bmatrix} \cdot \begin{bmatrix} -3 \\ 6 \\ -3 \end{bmatrix} = -12 + 30 - 18 = 0$$

7.6 Matrix Representation of Sets of Linear Equations

One reason for learning linear algebra concepts is to simplify representations for very complex systems. An engineering problem may involve hundreds or thousands of equations and variables. The same system can be represented simply as $A\,x = b$, where the vector x is a vector of unknowns and the values in the matrix A and vector b define the problem.

Example 7.8 (Linear Equations in Matrix Vector Form) Consider again the equations from the original distillation column example:

$$\begin{aligned} x &= 60 \\ y - x &= 30 \\ z - y &= -20 \end{aligned}$$

Recast these equations, writing each equation in the same order. Include values of 0 or 1 as coefficients where needed.

$$\begin{aligned} 1x + 0y + 0z &= 60 \\ -1x + 1y + 0z &= 30 \\ 0x - 1y + 1z &= -20 \end{aligned}$$

Notice that the variables (with constant coefficients) are on the left hand side and constant values are on the right hand side. This set of linear equations can be represented in the compact notation $A\,x = b$, where:

$$A = \begin{bmatrix} 1 & 0 & 0 \\ -1 & 1 & 0 \\ 0 & -1 & 1 \end{bmatrix}$$

$$x = \begin{bmatrix} x \\ y \\ z \end{bmatrix}$$

$$b = \begin{bmatrix} 60 \\ 30 \\ -20 \end{bmatrix}$$

7.6.1   Solving Sets of Linear Equations

We need a solution to the matrix equation $\underline{\underline{A}}\,\underline{x} = \underline{b}$. You cannot "divide" by a matrix:

$$\underline{x} \neq \underline{b}/\underline{\underline{A}}$$

There is no standard "division" operator for a matrix. Instead, an inverse is defined for some square matrices such that

$$\underline{\underline{A}} \left(\underline{\underline{A}}\right)^{-1} = \underline{\underline{I}}$$

Also, $\underline{\underline{A}}$ and $\underline{\underline{A}}^{-1}$ (if they exist) are commutable:

$$\left(\underline{\underline{A}}\right)^{-1} \underline{\underline{A}} = \underline{\underline{I}}$$

Now, there are just a few steps required to solve $\underline{\underline{A}}\underline{x} = \underline{b}$ for $\underline{x}$. First, multiply $\left(\underline{\underline{A}}\right)^{-1}$ on the left of both sides of the equation, $\underline{\underline{A}}\,\underline{x} = \underline{b}$ such that:

$$\left(\underline{\underline{A}}\right)^{-1} \underline{\underline{A}}\underline{x} = \left(\underline{\underline{A}}\right)^{-1} \underline{b}$$

Realizing that $\left(\underline{\underline{A}}\right)^{-1} \underline{\underline{A}} = \underline{\underline{I}}$, replace $\left(\underline{\underline{A}}\right)^{-1} \underline{\underline{A}}$ with $\underline{\underline{I}}$.

$$\underline{\underline{I}}\underline{x} = \left(\underline{\underline{A}}\right)^{-1} \underline{b}$$

Now, realizing $\underline{\underline{I}}\,\underline{x}$ is $\underline{x}$, the solution is:

$$\underline{x} = \left(\underline{\underline{A}}\right)^{-1} \underline{b}$$

Note that multiplying on the right will not lead to a solution.

$$\underline{\underline{A}}\underline{x} \left(\underline{\underline{A}}\right)^{-1} = \underline{b} \left(\underline{\underline{A}}\right)^{-1}$$

7.6.2 Determining the Matrix Inverse

To solve $\underline{\underline{A}}\,\underline{x} = \underline{b}$, you need to know $\left(\underline{\underline{A}}\right)^{-1}$. We are going to use row reduction to calculate $\left(\underline{\underline{A}}\right)^{-1}$. Start with $\underline{\underline{A}} \mid \underline{\underline{I}}$. Use row reduction techniques until $\underline{\underline{A}}$ is $\underline{\underline{I}}$. Matrix $\left(\underline{\underline{A}}\right)^{-1}$ (if it exists) will be on the right side where matrix $\underline{\underline{I}}$ was originally.

Example 7.9 Inverse Matrix Example

Solve the following for $\underline{x}$ using $\left(\underline{\underline{A}}\right)^{-1}$:

$$\begin{bmatrix} 1 & 2 \\ 3 & 4 \end{bmatrix} \underline{x} = \begin{bmatrix} 5 \\ 6 \end{bmatrix}$$

For this procedure, one must first calculate $\left(\underline{\underline{A}}\right)^{-1}$. Set up $\underline{\underline{A}} \mid \underline{\underline{I}}$ as:

$$\begin{array}{cc|cc} 1 & 2 & 1 & 0 \\ 3 & 4 & 0 & 1 \end{array}$$

Use row reduction to get

$$\begin{array}{cc|cc} 1 & 0 & ? & ? \\ 0 & 1 & ? & ? \end{array}$$

Then, verify that $\underline{\underline{A}}\left(\underline{\underline{A}}\right)^{-1} = \underline{\underline{I}}$. Use $\left(\underline{\underline{A}}\right)^{-1}$ to calculate $\underline{x}$ using $\underline{x} = \left(\underline{\underline{A}}\right)^{-1}\underline{b}$. Verify solution again to be safe.

START

Start by using row reduction on:

$$\begin{array}{cc|cc} 1 & 2 & 1 & 0 \\ 3 & 4 & 0 & 1 \end{array}$$

Multiply row 2 by 1/3 to get:

$$\begin{array}{cc|cc} 1 & 2 & 1 & 0 \\ 1 & \frac{4}{3} & 0 & \frac{1}{3} \end{array}$$

Then, subtract row 1 from row 2 to get:

$$\begin{array}{cc|cc} 1 & 2 & 1 & 0 \\ 0 & -\frac{2}{3} & -1 & \frac{1}{3} \end{array}$$

Now, multiply row 2 by $-3/2$ to get:

$$\begin{array}{cc|cc} 1 & 2 & 1 & 0 \\ 0 & 1 & \frac{3}{2} & -\frac{1}{2} \end{array}$$

To get the left side looking like the identity matrix, subtract 2 times row 2 from row 1. Note that this is a compound use of row reduction rules.

$$\begin{array}{cc|cc} 1 & 0 & -2 & 1 \\ 0 & 1 & \frac{3}{2} & -\frac{1}{2} \end{array}$$

You now have $\underline{\underline{A}}^{-1} = \begin{bmatrix} -2 & 1 \\ \frac{3}{2} & -\frac{1}{2} \end{bmatrix}$.

Now, verify that $\underline{\underline{A}} \left(\underline{\underline{A}}\right)^{-1} = \underline{\underline{I}}$:

$$\begin{bmatrix} 1 & 2 \\ 3 & 4 \end{bmatrix} \begin{bmatrix} -2 & 1 \\ \frac{3}{2} & -\frac{1}{2} \end{bmatrix} = \begin{bmatrix} 1(-2)+2(\frac{3}{2}) & 1(1)+2(-\frac{1}{2}) \\ 3(-2)+4(\frac{3}{2}) & 3(1)+4(-\frac{1}{2}) \end{bmatrix} = \begin{bmatrix} 1 & 0 \\ 0 & 1 \end{bmatrix}$$

You may also verify that $\left(\underline{\underline{A}}\right)^{-1} \underline{\underline{A}} = \underline{\underline{I}}$:

$$\begin{bmatrix} -2 & 1 \\ \frac{3}{2} & -\frac{1}{2} \end{bmatrix} \begin{bmatrix} 1 & 2 \\ 3 & 4 \end{bmatrix} = \begin{bmatrix} -2+3 & -4+4 \\ \frac{3}{2}-\frac{3}{2} & 3-2 \end{bmatrix} = \begin{bmatrix} 1 & 0 \\ 0 & 1 \end{bmatrix}$$

Now, compute the solution, $\underline{x} = \left(\underline{\underline{A}}\right)^{-1} \underline{b}$.

$$\underline{x} = \begin{bmatrix} -2 & 1 \\ \frac{3}{2} & -\frac{1}{2} \end{bmatrix} \begin{bmatrix} 5 \\ 6 \end{bmatrix} = \begin{bmatrix} -10+6 \\ \frac{15}{2}-3 \end{bmatrix} = \begin{bmatrix} -4 \\ 4\frac{1}{2} \end{bmatrix}$$

Again, verify that the solution is the solution to the original equations:

$$\begin{bmatrix} 1 & 2 \\ 3 & 4 \end{bmatrix} \underline{x} = \begin{bmatrix} 5 \\ 6 \end{bmatrix}$$

$$\begin{bmatrix} 1 & 2 \\ 3 & 4 \end{bmatrix} \begin{bmatrix} -4 \\ 4\frac{1}{2} \end{bmatrix} = \begin{bmatrix} -4+9 \\ -12+18 \end{bmatrix} = \begin{bmatrix} 5 \\ 6 \end{bmatrix}$$

Just as expected...

Example 7.10 Steady-State Simulation Example

Consider the cell model from Example 4.4. This model involves three species and two reactions:

$$V \frac{dC_A}{dt}(t) = D_A (C_{Ao} - C_A(t)) - V k_1 C_A(t)$$

$$V \frac{dC_B}{dt}(t) = V k_1 C_A(t) - V k_2 C_B(t)$$

$$V \frac{dC_C}{dt}(t) = V k_2 C_B(t) - D_C (C_C(t) - C_{Co}))$$

If the system is at steady-state, the time derivatives all become 0.

$$0 = D_A (C_{Ao} - C_A(t)) - V k_1 C_A(t)$$

$$0 = V k_1 C_A(t) - V k_2 C_B(t)$$

$$0 = V k_2 C_B(t) - D_C (C_C(t) - C_{Co}))$$

Assume that the concentrations inside and outside the cell are known. Let $C_A = 80 \frac{mmol}{L}$, $C_{Ao} = 100 \frac{mmol}{L}$, $C_B = 40 \frac{mmol}{L}$, $C_C = 20 \frac{mmol}{L}$, and $C_{Co} = 10 \frac{mmol}{L}$. The equations become:

$$0 = D_A (100 - 80) - V k_1 80$$

$$0 = V k_1 80 - V k_2 40$$

$$0 = V k_2 40 - D_C (20 - 10))$$

These three equations leave us with five variables: D_A, V, k_1, k_2, and D_C. Three algebraic equations cannot provide enough information to solve for five unknown variables. Either more equations must be derived or some of the variables must be assigned values. Assume that the effective diffusion coefficient is approximated as $D_A = 2$. Also, assume that the cell volume may be approximated as $100\,L$. This leaves three unknowns, D_C, k_1, and k_2.

$$0 = 2 (100 - 80) - 100 k_1 80$$

$$0 = 100 k_1 80 - 100 k_2 40$$

$$0 = 100 k_2 40 - D_C (20 - 10))$$

The coefficients can be combined and the equations can be rearranged. Note that each equation is written in terms of all three variables, k_1, k_2, and D_C. Any constant terms are moved to the right hand side of the equation, while all variables are moved to the left hand side. Constant coefficients 0 and 1 can be included in front of every variable as well.

$$8000 k_1 + 0 k_2 + 0 D_C = 40$$

$$100 k_1 - 100 k_2 + 0 D_C = 0$$

$$0 k_1 + 400 k_2 - 10 D_C = 0$$

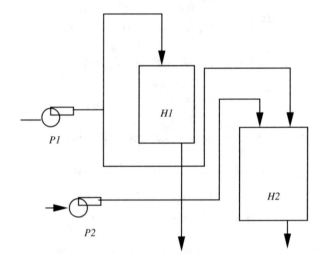

Fig. 7.2 Pump/tank example

These three equations can be put in the form of $Ax = b$ as:

$$\begin{bmatrix} 8000 & 0 & 0 \\ 100 & -100 & 0 \\ 0 & 400 & -10 \end{bmatrix} \begin{bmatrix} k_1 \\ k_2 \\ D_C \end{bmatrix} = \begin{bmatrix} 40 \\ 0 \\ 0 \end{bmatrix}$$

Example 7.11 Steady-State Control Example

Two pumps are used to fill two tanks. The pumps usually operate at 50%, keeping the tanks at levels of 75 inches and 80 inches, respectively. It is known that a 1% increase in pump 1 increases the height of tank 1 by 5 inches and the height of tank 2 by 3 inches. For a 1% change in pump 2, the height of tank 2 increases by 4 inches. It is desired to change the operating levels of the tanks to 110 inches and 89 inches (Fig. 7.2).

What do you know:

$$5\,\Delta P_1(\%) \qquad\qquad = \Delta H_1(inches)$$
$$3\,\Delta P_1(\%) + 4\,\Delta P_1(\%) = \Delta H_2(inches)$$

You know the target (reference, set point) for H_1 and H_2 as 110 and 89. This translates into $\Delta H_1 = 110 - 75 = 35$ and $\Delta H_2 = 89 - 80 = 9$. You need to increase tank 1 by 35 inches and tank 2 by 9 inches. You do not know the final values of the pump speeds. You do know the original steady-state values, 50% and 50%, realizing that:

$$P_{final} = P_{ss} + \Delta P$$

You can now set up linear equations to solve for ΔP_1 and ΔP_2, and then calculate the final values for the pump speeds.

$$\begin{bmatrix} 5 & 0 \\ 3 & 4 \end{bmatrix} \begin{bmatrix} \Delta P_1 \\ \Delta P_2 \end{bmatrix} = \begin{bmatrix} \Delta H_1 \\ \Delta H_2 \end{bmatrix}$$

7.7 Visualization

Each row in $\underline{A}\,\underline{x} = \underline{b}$ is a single linear equation. For a 2D problem ($\underline{x}$ with 2 elements/unknowns), the equation defines a line in the (x, y) plane. Two equations define two lines, and the unique solution to $\underline{A}\,\underline{x} = \underline{b}$ is the point $\underline{x}$ where the lines intersect. In some cases, there may be many solutions to $\underline{A}\,\underline{x} = \underline{b}$, and in some cases, there may be no solutions to $\underline{A}\,\underline{x} = \underline{b}$ (Fig. 7.3).

For a 3D problem, each row defines the equation for a plane in three-dimensional space. The intersection of two non-parallel planes is a line in three-dimensional space, and the intersection of a line and a plane in three-dimensional space is a point. Again, in some cases there may be a single solution, many solutions, or no solutions.

For higher dimensions, each equation defines a *hyperplane* in an n-dimensional space, $\mathbb{R}^n$.

7.7.1 Linear Transform

A vector in $\mathbb{R}^n$ means that x has n real-valued elements. Matrix multiplication of a matrix of size $m \times n$ times a vector of size $n \times 1$ "maps" the vector from $\mathbb{R}^n$ to $\mathbb{R}^m$ (Fig. 7.4).

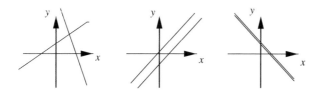

Fig. 7.3 Three 2D examples with two equations. Each equation (row) represents a line. The first case has one solution, the second case has no solution, and the third case has many solutions

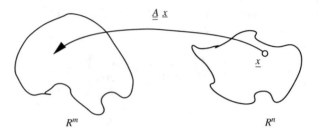

Fig. 7.4 Matrix multiplication as a mapping from $\mathbb{R}^n$ to $\mathbb{R}^m$ for $m \times n$ matrix $\underline{\underline{A}}$

7.7.2 Range

The range of a matrix is the space of all possible points that may be mapped to in a matrix multiplication of that matrix times an unknown vector.

Example 7.12 Range Example 1

For example, the matrix:

$$\underline{\underline{A}} = \begin{bmatrix} 1 & 1 & 0 \\ 1 & 1 & 0 \\ 0 & 0 & 0 \end{bmatrix}$$

can only map to points on the line $x + y$ in 3D as follows:

$$\underline{\underline{A}}\,x = 2x + 2y + 0z$$

The columns of the matrix define possible directions for the matrix to transform a vector. In this example, columns 1 and 2 are the same, and column 3 is the zero vector. $\underline{\underline{A}}\,x$ where x takes any real value will always be on the line defined by the direction $\begin{bmatrix} 1 \\ 1 \\ 0 \end{bmatrix}$.

Example 7.13 Range Example 2

In another example, the matrix:

$$\underline{\underline{A}} = \begin{bmatrix} 1 & 0 & 0 \\ 1 & 1 & 0 \\ 0 & 0 & 0 \end{bmatrix}$$

can only map to a variety of points in 3D as follows:

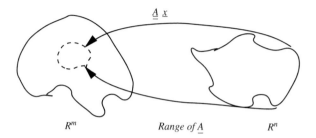

$$R^m \qquad \textit{Range of } \underline{A} \qquad R^n$$

Fig. 7.5 Range of $\underline{A}$ as space in $\mathbb{R}^m$ of all possible mappings from $\mathbb{R}^n$ using matrix multiplication

$$\underline{A}\,\underline{x} = \begin{bmatrix} 1 \\ 1 \\ 0 \end{bmatrix} x + \begin{bmatrix} 0 \\ 1 \\ 0 \end{bmatrix} y + \begin{bmatrix} 0 \\ 0 \\ 0 \end{bmatrix} z$$

Again, the columns of the matrix define possible directions for the matrix to transform a vector. In this example, only points in the directions of $\begin{bmatrix} 1 \\ 1 \\ 0 \end{bmatrix}$ and $\begin{bmatrix} 0 \\ 1 \\ 0 \end{bmatrix}$ can be reached when multiplying $\underline{A}\,\underline{x}$. These two directions form a plane in three-dimensional space (Fig. 7.5).

Example 7.14 Range Example 3

In another example, the matrix:

$$\underline{A} = \begin{bmatrix} 1 & 0 & 1 \\ 1 & 1 & 2 \\ 0 & 0 & 0 \end{bmatrix}$$

can only map to a variety of points in 3D as follows:

$$\underline{A}\,\underline{x} = \begin{bmatrix} 1 \\ 1 \\ 0 \end{bmatrix} x + \begin{bmatrix} 0 \\ 1 \\ 0 \end{bmatrix} y + \begin{bmatrix} 0 \\ 0 \\ 0 \end{bmatrix} z$$

Here, column 3 is linearly dependent upon columns 1 and 2. This means that you can find some combination of columns 1 and 2 that give column 3. Column 3 lies in the plane defined by columns 1 and 2.

For $\underline{A}\,\underline{x} = \underline{b}$ to Have a Solution, the $\underline{b}$ Must Be in the Range of $\underline{A}$

For the last examples, if $\underline{b} = \begin{bmatrix} ? \\ ? \\ 1 \end{bmatrix}$ (if $\underline{b}$ has element in the z position), there will not

be a solution to $\underline{\underline{A}}\,\underline{x} = \underline{b}$. In such a case, the possible range of $\underline{\underline{A}}$ does not include $\underline{b}$.

Example 7.15 Range Example 4

In another example, the matrix:

$$\underline{\underline{A}} = \begin{bmatrix} 1 & 0 & 1 \\ 1 & 1 & 2 \\ 0 & 0 & 1 \end{bmatrix}$$

can map to all of the points in 3D as follows:

$$\underline{\underline{A}}\,\underline{x} = \begin{bmatrix} 1 \\ 1 \\ 0 \end{bmatrix} x + \begin{bmatrix} 0 \\ 1 \\ 0 \end{bmatrix} y + \begin{bmatrix} 1 \\ 2 \\ 1 \end{bmatrix} z$$

Here, column 3 is NOT linearly dependent upon columns 1 and 2. This means that you can find some combination of columns 1, 2, and 3 that give any point in three dimensions.

Rank The rank of a matrix is the number of linearly independent columns. For a square matrix of size $n \times n$, there is a unique solution if there are n independent columns. The matrix would have rank n.

7.8   Mass Balance Example

This lengthy example sets up model equations for a relatively simple steady-state reaction system. The equations are described and put into matrix form. Additionally, MATLAB code is presented to solve the same set of equations.

Example 7.16 Mass Balance Example

Consider two reactors involving three species. The two reactors operate at steady-state, so the accumulation rates are 0. In reactor 1, species A reacts with extent ξ_1 to form B and C. Also, in reactor 1, species B reacts to form C with extent ξ_2. In reactor 2, species A is converted to B with extent ξ_3. For each reactor and each species, the general form of the mass balance equation becomes:

$$ACCUMULATION = IN - OUT + CREATED - REACTED$$

Here, $ACCUMULATION$ is the molar accumulation rate in the unit, IN is the sum of molar flow rates into the unit, OUT is the total molar flow rate out of the unit,

$CREATED$ is the total molar reaction rate creating that species in the unit, and $REACTED$ is the total molar rate for consumption/conversion of that species in the unit. Assuming the reactor is at steady-state means $ACCUMULATION = 0$. This general mass balance equation can be converted to the form:

$$IN + CREATED - REACTED = OUT$$

The amount of moles of a species is often denoted using the variable n. Molar flow rates are often denoted using dot notation, resulting a variable $\dot{n}$. This type of variable can be given subscripts to denote different species flowing into and out of different units.

Molar flow rates for the two reactors and three species are specified as follows. The molar feed rates for A, B, and C to reactor 1 are $\dot{n}_{A0}$, $\dot{n}_{B0}$, and $\dot{n}_{C0}$, respectively. The molar rate flowing from reactor 1 and being fed to reactor 2 is given by the variables $\dot{n}_{A1}$, $\dot{n}_{B1}$, and $\dot{n}_{C1}$. The final product of molar flow rates from reactor 2 is given by variables $\dot{n}_{A2}$, $\dot{n}_{B2}$, and $\dot{n}_{C2}$. Assuming the reaction extents for the three reactions are ξ_1, ξ_2, and ξ_3, there are 12 total variables:

$$x = [\dot{n}_{A0} \ \dot{n}_{B0} \ \dot{n}_{C0} \ \dot{n}_{A1} \ \dot{n}_{B1} \ \dot{n}_{C1} \ \dot{n}_{A2} \ \dot{n}_{B2} \ \dot{n}_{C2} \ \xi_1 \ \xi_2 \ \xi_3]^T$$

The transpose operator converts a row vector to a column vector. All variables have units of $\left(\frac{mol}{time}\right)$, where time can be any appropriate time unit.

Assuming the feed to reactor 1 is $100 \ \frac{mol}{hr}$, a simple equation is produced:

$$\dot{n}_{A0} = 100$$

Writing this equation with all of the twelve variables results in:

$$1\dot{n}_{A0} + 0\dot{n}_{B0} + 0\dot{n}_{C0} + 0\dot{n}_{A1} + 0\dot{n}_{B1} + 0\dot{n}_{C1} + 0\dot{n}_{A2} + 0\dot{n}_{B2} + 0\dot{n}_{C2} + 0\xi_1 + 0\xi_2 + 0\xi_3 = 100$$

Given the vector of variables x, the left hand side can be put in matrix vector form:

$$[1\,0\,0\,0\,0\,0\,0\,0\,0\,0\,0\,0]\,x = 100$$

The left side is basically the dot product of two vectors. One vector is a row vector containing only zeros and a single coefficient of one. The other vector is a column vector containing twelve different variables. Specifying feed rates for B and C can also be accomplished with the following two equations:

$$\dot{n}_{B0} = 20$$
$$\dot{n}_{C0} = 5$$

These can also be written out with all the twelve variables as:

$$0\dot{n}_{A0} + 1\dot{n}_{B0} + 0\dot{n}_{C0} + 0\dot{n}_{A1} + 0\dot{n}_{B1} + 0\dot{n}_{C1} + 0\dot{n}_{A2} + 0\dot{n}_{B2} + 0\dot{n}_{C2} + 0\xi_1 + 0\xi_2 + 0\xi_3 = 20$$

$$0\dot{n}_{A0} + 0\dot{n}_{B0} + 1\dot{n}_{C0} + 0\dot{n}_{A1} + 0\dot{n}_{B1} + 0\dot{n}_{C1} + 0\dot{n}_{A2} + 0\dot{n}_{B2} + 0\dot{n}_{C2} + 0\xi_1 + 0\xi_2 + 0\xi_3 = 5$$

which can be represented in matrix vector form as:

$$\begin{bmatrix} 0 1 0 0 0 0 0 0 0 0 0 0 \\ 0 0 1 0 0 0 0 0 0 0 0 0 \end{bmatrix} x = \begin{bmatrix} 20 \\ 5 \end{bmatrix}$$

Outlet flow rates can similarly be specified with three more simple equations:

$$\dot{n}_{A2} = 50$$

$$\dot{n}_{B2} = 55$$

$$\dot{n}_{C2} = 50$$

Including all the variables for these three equations looks like;

$$0\dot{n}_{A0} + 0\dot{n}_{B0} + 0\dot{n}_{C0} + 0\dot{n}_{A1} + 0\dot{n}_{B1} + 0\dot{n}_{C1} + 1\dot{n}_{A2} + 0\dot{n}_{B2} + 0\dot{n}_{C2} + 0\xi_1 + 0\xi_2 + 0\xi_3 = 50$$

$$0\dot{n}_{A0} + 0\dot{n}_{B0} + 0\dot{n}_{C0} + 0\dot{n}_{A1} + 0\dot{n}_{B1} + 0\dot{n}_{C1} + 0\dot{n}_{A2} + 1\dot{n}_{B2} + 0\dot{n}_{C2} + 0\xi_1 + 0\xi_2 + 0\xi_3 = 55$$

$$0\dot{n}_{A0} + 0\dot{n}_{B0} + 0\dot{n}_{C0} + 0\dot{n}_{A1} + 0\dot{n}_{B1} + 0\dot{n}_{C1} + 0\dot{n}_{A2} + 0\dot{n}_{B2} + 1\dot{n}_{C2} + 0\xi_1 + 0\xi_2 + 0\xi_3 = 50$$

The overall system of six equations now in the $Ax = b$ form now becomes:

$$\begin{bmatrix} 1 0 0 0 0 0 0 0 0 0 0 0 \\ 0 1 0 0 0 0 0 0 0 0 0 0 \\ 0 0 1 0 0 0 0 0 0 0 0 0 \\ 0 0 0 0 0 0 1 0 0 0 0 0 \\ 0 0 0 0 0 0 0 1 0 0 0 0 \\ 0 0 0 0 0 0 0 0 1 0 0 0 \end{bmatrix} \begin{bmatrix} \dot{n}_{A0} \\ \dot{n}_{B0} \\ \dot{n}_{C0} \\ \dot{n}_{A1} \\ \dot{n}_{B1} \\ \dot{n}_{AC1} \\ \dot{n}_{A2} \\ \dot{n}_{B2} \\ \dot{n}_{C2} \\ \xi_1 \\ \xi_2 \\ \xi_3 \end{bmatrix} = \begin{bmatrix} 100 \\ 20 \\ 5 \\ 50 \\ 55 \\ 50 \end{bmatrix}$$

where the matrix A and the vector b are defined as:

$$A = \begin{bmatrix} 1 & 0 & 0 & 0 & 0 & 0 & 0 & 0 & 0 & 0 & 0 & 0 \\ 0 & 1 & 0 & 0 & 0 & 0 & 0 & 0 & 0 & 0 & 0 & 0 \\ 0 & 0 & 1 & 0 & 0 & 0 & 0 & 0 & 0 & 0 & 0 & 0 \\ 0 & 0 & 0 & 0 & 0 & 0 & 1 & 0 & 0 & 0 & 0 & 0 \\ 0 & 0 & 0 & 0 & 0 & 0 & 0 & 1 & 0 & 0 & 0 & 0 \\ 0 & 0 & 0 & 0 & 0 & 0 & 0 & 0 & 1 & 0 & 0 & 0 \end{bmatrix} \quad b = \begin{bmatrix} 100 \\ 20 \\ 5 \\ 50 \\ 55 \\ 50 \end{bmatrix}$$

Note that in matrix A, columns 4, 5, and 6 only include values of 0. This means that variables $\dot{n}_{A1}$, $\dot{n}_{B1}$, and $\dot{n}_{C1}$ are not involved in any way in these equations. Similarly, columns 10, 11, and 12 have only 0 in each entry. The reaction extent variable values are not being used in these six equations.

The mass balance equation for species A in reactor 1 involves the molar reaction rate for reaction 1, ξ_1:

$$\dot{n}_{A0} - \xi_1 = \dot{n}_{A1}$$

Getting everything on the left side of the equation and writing all the variables leads to:

$$1\dot{n}_{A0} + 0\dot{n}_{B0} + 0\dot{n}_{C0} - 1\dot{n}_{A1} + 0\dot{n}_{B1} + 0\dot{n}_{C1} + 0\dot{n}_{A2} + 0\dot{n}_{B2} + 0\dot{n}_{C2} - 1\xi_1 + 0\xi_2 + 0\xi_3 = 0$$

For B, the balance involves two reactions. Species B is created by reaction 1 and it is consumed in species 2. The balance is:

$$\dot{n}_{B0} + \xi_1 - \xi_2 = \dot{n}_{B1}$$

Or with all the variables on the left side,

$$0\dot{n}_{A0} + 1\dot{n}_{B0} + 0\dot{n}_{C0} + 0\dot{n}_{A1} - 1\dot{n}_{B1} + 0\dot{n}_{C1} + 0\dot{n}_{A2} + 0\dot{n}_{B2} + 0\dot{n}_{C2} + 1\xi_1 - 1\xi_2 + 0\xi_3 = 0$$

For C, both reactions create the species in the reactor:

$$\dot{n}_{C0} + \xi_1 + \xi_2 = \dot{n}_{C1}$$

Writing this equation in terms of all the variables becomes:

$$0\dot{n}_{A0} + 0\dot{n}_{B0} + 1\dot{n}_{C0} + 0\dot{n}_{A1} + 0\dot{n}_{B1} - 1\dot{n}_{C1} + 0\dot{n}_{A2} + 0\dot{n}_{B2} + 0\dot{n}_{C2} + 1\xi_1 + 1\xi_2 + 0\xi_3 = 0$$

Now, adding these three equations, the A matrix and b vector representing the full set of equations in $Ax = b$ form are:

$$A = \begin{bmatrix} 1 & 0 & 0 & 0 & 0 & 0 & 0 & 0 & 0 & 0 & 0 & 0 \\ 0 & 1 & 0 & 0 & 0 & 0 & 0 & 0 & 0 & 0 & 0 & 0 \\ 0 & 0 & 1 & 0 & 0 & 0 & 0 & 0 & 0 & 0 & 0 & 0 \\ 0 & 0 & 0 & 0 & 0 & 0 & 1 & 0 & 0 & 0 & 0 & 0 \\ 0 & 0 & 0 & 0 & 0 & 0 & 0 & 1 & 0 & 0 & 0 & 0 \\ 0 & 0 & 0 & 0 & 0 & 0 & 0 & 0 & 1 & 0 & 0 & 0 \\ 1 & 0 & 0 & -1 & 0 & 0 & 0 & 0 & -1 & 0 & 0 & 0 \\ 0 & 1 & 0 & 0 & -1 & 0 & 0 & 0 & 0 & 1 & -1 & 0 \\ 0 & 0 & 1 & 0 & 0 & -1 & 0 & 0 & 0 & 1 & 1 & 0 \end{bmatrix} \quad b = \begin{bmatrix} 100 \\ 20 \\ 5 \\ 50 \\ 55 \\ 50 \\ 0 \\ 0 \\ 0 \end{bmatrix}$$

Moving on to reactor 2, we remember that reactor 2 only has one reaction involving A and B

$$\dot{n}_{A1} - \xi_3 = \dot{n}_{A2}$$

$$\dot{n}_{B1} + \xi_3 = \dot{n}_{B2}$$

$$\dot{n}_{C1} = \dot{n}_{C2}$$

Writing these three equations in terms of all the variables:

$$0\dot{n}_{A0} + 0\dot{n}_{B0} + 0\dot{n}_{C0} + 1\dot{n}_{A1} + 0\dot{n}_{B1} + 0\dot{n}_{C1} - 1\dot{n}_{A2} + 0\dot{n}_{B2} + 0\dot{n}_{C2} + 0\xi_1 + 0\xi_2 - 1\xi_3 = 0$$

$$0\dot{n}_{A0} + 0\dot{n}_{B0} + 0\dot{n}_{C0} + 0\dot{n}_{A1} + 1\dot{n}_{B1} + 0\dot{n}_{C1} + 0\dot{n}_{A2} - 1\dot{n}_{B2} + 0\dot{n}_{C2} + 0\xi_1 + 0\xi_2 + 1\xi_3 = 0$$

$$0\dot{n}_{A0} + 0\dot{n}_{B0} + 0\dot{n}_{C0} + 0\dot{n}_{A1} + 0\dot{n}_{B1} + 1\dot{n}_{C1} + 0\dot{n}_{A2} + 0\dot{n}_{B2} - 1\dot{n}_{C2} + 0\xi_1 + 0\xi_2 + 0\xi_3 = 0$$

The resulting A matrix and b vector become:

$$A = \begin{bmatrix} 1 & 0 & 0 & 0 & 0 & 0 & 0 & 0 & 0 & 0 & 0 & 0 \\ 0 & 1 & 0 & 0 & 0 & 0 & 0 & 0 & 0 & 0 & 0 & 0 \\ 0 & 0 & 1 & 0 & 0 & 0 & 0 & 0 & 0 & 0 & 0 & 0 \\ 0 & 0 & 0 & 0 & 0 & 0 & 1 & 0 & 0 & 0 & 0 & 0 \\ 0 & 0 & 0 & 0 & 0 & 0 & 0 & 1 & 0 & 0 & 0 & 0 \\ 0 & 0 & 0 & 0 & 0 & 0 & 0 & 0 & 1 & 0 & 0 & 0 \\ 1 & 0 & 0 & -1 & 0 & 0 & 0 & 0 & 0 & -1 & 0 & 0 \\ 0 & 1 & 0 & 0 & -1 & 0 & 0 & 0 & 0 & 1 & -1 & 0 \\ 0 & 0 & 1 & 0 & 0 & -1 & 0 & 0 & 0 & 1 & 1 & 0 \\ 0 & 0 & 0 & 1 & 0 & 0 & -1 & 0 & 0 & 0 & 0 & -1 \\ 0 & 0 & 0 & 0 & 1 & 0 & 0 & -1 & 0 & 0 & 0 & 1 \\ 0 & 0 & 0 & 0 & 0 & 1 & 0 & 0 & -1 & 0 & 0 & 0 \end{bmatrix} \quad b = \begin{bmatrix} 100 \\ 20 \\ 5 \\ 50 \\ 55 \\ 50 \\ 0 \\ 0 \\ 0 \\ 0 \\ 0 \\ 0 \end{bmatrix}$$

The system is currently set up with adequate equations and could be solved. There are 12 independent equations with 12 unknowns. They can be verified by checking the rank of the matrix. For a 12x12 matrix, there is a unique solution if the rank of the matrix is 12. This means that all the equations are independent. It also means that all of the columns of matrix A are linearly independent; they point in different directions in a 12-dimensional space. This also means that the determinant of the matrix should be nonzero, letting one know that the inverse matrix could be calculated. The determinant of this 12x12 matrix is -1.

However, overall mass balances can also be written for the two-reactor system as a whole:

$$\dot{n}_{A0} - \xi_1 - \xi_3 = \dot{n}_{A2}$$

$$\dot{n}_{B0} + \xi_1 - \xi_2 + \xi_3 = \dot{n}_{B2}$$

$$\dot{n}_{C0} + \xi_1 + \xi_2 = \dot{n}_{C2}$$

$$1\dot{n}_{A0} + 0\dot{n}_{B0} + 0\dot{n}_{C0} + 0\dot{n}_{A1} + 0\dot{n}_{B1} + 0\dot{n}_{C1} - 1\dot{n}_{A2} + 0\dot{n}_{B2} + 0\dot{n}_{C2} - 1\xi_1 + 0\xi_2 - 1\xi_3 = 0$$

$$0\dot{n}_{A0} + 1\dot{n}_{B0} + 0\dot{n}_{C0} + 0\dot{n}_{A1} + 0\dot{n}_{B1} + 0\dot{n}_{C1} + 0\dot{n}_{A2} - 1\dot{n}_{B2} + 0\dot{n}_{C2} + 1\xi_1 - 1\xi_2 - 1\xi_3 = 0$$

$$0\dot{n}_{A0} + 0\dot{n}_{B0} + 1\dot{n}_{C0} + 0\dot{n}_{A1} + 0\dot{n}_{B1} + 0\dot{n}_{C1} + 0\dot{n}_{A2} + 0\dot{n}_{B2} - 1\dot{n}_{C2} + 1\xi_1 + 1\xi_2 + 0\xi_3 = 0$$

These three new equations add three more rows onto A and three values onto b.

$$
A =
\begin{bmatrix}
1 & 0 & 0 & 0 & 0 & 0 & 0 & 0 & 0 & 0 & 0 & 0 \\
0 & 1 & 0 & 0 & 0 & 0 & 0 & 0 & 0 & 0 & 0 & 0 \\
0 & 0 & 1 & 0 & 0 & 0 & 0 & 0 & 0 & 0 & 0 & 0 \\
0 & 0 & 0 & 0 & 0 & 0 & 1 & 0 & 0 & 0 & 0 & 0 \\
0 & 0 & 0 & 0 & 0 & 0 & 0 & 1 & 0 & 0 & 0 & 0 \\
0 & 0 & 0 & 0 & 0 & 0 & 0 & 0 & 1 & 0 & 0 & 0 \\
1 & 0 & 0 & -1 & 0 & 0 & 0 & 0 & 0 & -1 & 0 & 0 \\
0 & 1 & 0 & 0 & -1 & 0 & 0 & 0 & 0 & 1 & -1 & 0 \\
0 & 0 & 1 & 0 & 0 & -1 & 0 & 0 & 0 & 1 & 1 & 0 \\
0 & 0 & 0 & 1 & 0 & 0 & -1 & 0 & 0 & 0 & 0 & -1 \\
0 & 0 & 0 & 0 & 1 & 0 & 0 & -1 & 0 & 0 & 0 & 1 \\
0 & 0 & 0 & 0 & 0 & 1 & 0 & 0 & -1 & 0 & 0 & 0 \\
1 & 0 & 0 & 0 & 0 & 0 & -1 & 0 & 0 & -1 & 0 & -1 \\
0 & 1 & 0 & 0 & 0 & 0 & 0 & -1 & 0 & 1 & -1 & 1 \\
0 & 0 & 1 & 0 & 0 & 0 & 0 & 0 & -1 & 1 & 1 & 0
\end{bmatrix}
\quad
b =
\begin{bmatrix}
100 \\
20 \\
5 \\
50 \\
55 \\
50 \\
0 \\
0 \\
0 \\
0 \\
0 \\
0 \\
0 \\
0 \\
0
\end{bmatrix}
$$

Now, these 15 equations together are not linearly independent. However, it is not easy to see this at first. The column rank of A^T is 12; this means that 12 of the 15 equations could be used to solve for all the unknowns. It is not always clear what equations are not needed. However, if you remove a single equation and the rank decreases, you know that the equation is necessary for solution of the overall set of equations. In this case, there are a variety of ways one could pick 12 of the 15 equations to get to a solution. However, the first six equations are required to be included; these equations specify the values of six of the variables explicitly for the reactor 1 feed and reactor 2 product streams. Similarly, you cannot remove the three equations for a single species simultaneously.

These six equations are explicit in one variable, since the feed to reactor 1 and the flow from reactor 2 are defined. These six variables can be removed to simplify the overall set of equations. Consider the three sets of equations (reactor 1, reactor 2, and overall) for reactor 1:

$$\dot{n}_{A0} - \xi_1 \qquad = \dot{n}_{A1}$$
$$\dot{n}_{B0} + \xi_1 - \xi_2 = \dot{n}_{B1}$$
$$\dot{n}_{C0} + \xi_1 + \xi_2 = \dot{n}_{C1}$$

Reactor 2:

$$\dot{n}_{A1} - \xi_3 = \dot{n}_{A2}$$
$$\dot{n}_{B1} + \xi_3 = \dot{n}_{B2}$$
$$\dot{n}_{C1} \qquad = \dot{n}_{C2}$$

Overall:

$$\dot{n}_{A0} - \xi_1 \qquad - \xi_3 = \dot{n}_{A2}$$
$$\dot{n}_{B0} + \xi_1 - \xi_2 + \xi_3 = \dot{n}_{B2}$$
$$\dot{n}_{C0} + \xi_1 + \xi_2 \qquad = \dot{n}_{C2}$$

Replacing the known variables for entering reactor 1 and leaving reactor 2 leaves us with the following. Reactor 1:

$$100 - \xi_1 = \dot{n}_{A1}$$
$$20 + \xi_1 - \xi_2 = \dot{n}_{B1}$$
$$5 + \xi_1 + \xi_2 = \dot{n}_{C1}$$

Reactor 2:

$$\dot{n}_{A1} - \xi_3 = 50$$
$$\dot{n}_{B1} + \xi_3 = 55$$
$$\dot{n}_{C1} = 50$$

Overall:

$$100 - \xi_1 - \xi_3 = 50$$
$$20 + \xi_1 - \xi_2 + \xi_3 = 55$$
$$5 + \xi_1 + \xi_2 = 50$$

Note that, of these nine equations, only one of them is explicit in a single variable. There are six variables and nine possible equations. Writing out the mass balance equations including all variables results in the following for reactor 1:

$$-1\dot{n}_{A1} + 0\dot{n}_{B1} + 0\dot{n}_{C1} - 1\xi_1 + 0\xi_2 + 0\xi_3 = -100$$
$$0\dot{n}_{A1} - 1\dot{n}_{B1} + 0\dot{n}_{C1} + 1\xi_1 - 1\xi_2 + 0\xi_3 = -20$$
$$0\dot{n}_{A1} + 0\dot{n}_{B1} - 1\dot{n}_{C1} + 1\xi_1 + 1\xi_2 + 0\xi_3 = -5$$

Reactor 2:

$$1\dot{n}_{A1} + 0\dot{n}_{B1} + 0\dot{n}_{C1} + 0\xi_1 + 0\xi_2 - 1\xi_3 = 50$$
$$0\dot{n}_{A1} + 1\dot{n}_{B1} + 0\dot{n}_{C1} + 0\xi_1 + 0\xi_2 + 1\xi_3 = 55$$
$$0\dot{n}_{A1} + 0\dot{n}_{B1} + 1\dot{n}_{C1} + 0\xi_1 + 0\xi_2 + 0\xi_3 = 50$$

Overall:

$$0\dot{n}_{A1} + 0\dot{n}_{B1} + 0\dot{n}_{C1} - 1\xi_1 + 0\xi_2 + 0\xi_3 = -50$$
$$0\dot{n}_{A1} + 0\dot{n}_{B1} + 0\dot{n}_{C1} + 1\xi_1 - 1\xi_2 + 0\xi_3 = 35$$
$$0\dot{n}_{A1} + 0\dot{n}_{B1} + 0\dot{n}_{C1} + 1\xi_1 + 1\xi_2 + 0\xi_3 = 45$$

Now, it may be a bit more obvious that adding the reactor 1 equations to the reactor 2 equations gives the overall balance. There are six variables and nine equations. In the matrix form,

$$
A = \begin{bmatrix}
-1 & 0 & 0 & -1 & 0 & 0 \\
0 & -1 & 0 & 1 & -1 & 0 \\
0 & 0 & -1 & 1 & 1 & 0 \\
1 & 0 & 0 & 0 & 0 & -1 \\
0 & 1 & 0 & 0 & 0 & 1 \\
0 & 0 & 1 & 0 & 0 & 0 \\
0 & 0 & 0 & -1 & 0 & -1 \\
0 & 0 & 0 & 1 & -1 & 1 \\
0 & 0 & 0 & 1 & 1 & 0
\end{bmatrix}
\quad b = \begin{bmatrix}
-100 \\
-20 \\
-5 \\
50 \\
55 \\
50 \\
-50 \\
35 \\
45
\end{bmatrix}
$$

Adding overall balances when you already have balances for each unit generally leads to an over-specified set of equations. Therefore, consider only reactors 1 and 2. The six equations are:

$$100 - \xi_1 = \dot{n}_{A1}$$

$$20 + \xi_1 - \xi_2 = \dot{n}_{B1}$$

$$5 + \xi_1 + \xi_2 = \dot{n}_{C1}$$

$$\dot{n}_{A1} - \xi_3 = 50$$

$$\dot{n}_{B1} + \xi_3 = 55$$

$$\dot{n}_{C1} = 50$$

Note again that every equation except one includes at least two variables. Even replacing $\dot{n}_{C1}$ with a value of 50 in the mass balance for reactor 1 still leaves two variables in the third equation. In the matrix form, this becomes:

$$A = \begin{bmatrix} -1 & 0 & 0 & -1 & 0 & 0 \\ 0 & -1 & 0 & 1 & -1 & 0 \\ 0 & 0 & -1 & 1 & 1 & 0 \\ 1 & 0 & 0 & 0 & 0 & -1 \\ 0 & 1 & 0 & 0 & 0 & 1 \\ 0 & 0 & 1 & 0 & 0 & 0 \end{bmatrix} \quad b = \begin{bmatrix} -100 \\ -20 \\ -5 \\ 50 \\ 55 \\ 50 \end{bmatrix}$$

The rank of the 6x6 matrix A is 6, meaning that it is full rank and that a unique solution exists. The determinant of A is 1, so the matrix is invertible. The system of equations can be solved using the matrix inverse A^{-1} and the formula:

$$x = A^{-1}b$$

yielding a solution:

$$x = \begin{bmatrix} 70 & 35 & 50 & 30 & 15 & 20 \end{bmatrix}^T$$

Again, variable $\dot{n}_{C1}$ is explicitly known to be 50 since we have the equation $\dot{n}_{C1} = 50$. This means that the variable and the corresponding equation can be easily removed, resulting in the following five equations:

$$100 - \xi_1 = \dot{n}_{A1}$$

$$20 + \xi_1 - \xi_2 = \dot{n}_{B1}$$

$$5 + \xi_1 + \xi_2 = 50$$

$$\dot{n}_{A1} - \xi_3 = 50$$

$$\dot{n}_{B1} + \xi_3 = 55$$

Writing the equations in terms of all five variables and moving the constants to the right hand side results in:

$$-1\dot{n}_{A1} + 0\dot{n}_{B1} - 1\xi_1 + 0\xi_2 + 0\xi_3 = -100$$

$$0\dot{n}_{A1} - 1\dot{n}_{B1} + 1\xi_1 - 1\xi_2 + 0\xi_3 = -20$$

$$0\dot{n}_{A1} + 0\dot{n}_{B1} + 1\xi_1 + 1\xi_2 + 0\xi_3 = 45$$

$$1\dot{n}_{A1} + 0\dot{n}_{B1} + 0\xi_1 + 0\xi_2 - 1\xi_3 = 50$$

$$0\dot{n}_{A1} + 1\dot{n}_{B1} + 0\xi_1 + 0\xi_2 + 1\xi_3 = 55$$

The resulting five equations with five unknowns are represented in $Ax = b$ form as:

$$A = \begin{bmatrix} -1 & 0 & -1 & 0 & 0 \\ 0 & -1 & 1 & -1 & 0 \\ 0 & 0 & 1 & 1 & 0 \\ 1 & 0 & 0 & 0 & -1 \\ 0 & 1 & 0 & 0 & 1 \end{bmatrix} \quad b = \begin{bmatrix} -100 \\ -20 \\ 45 \\ 50 \\ 55 \end{bmatrix}$$

Note that each row of matrix A has at least two nonzero values. This means that each equation includes at least two variables. There is no way to use any equation to solve directly for a variable. This system of equations requires solution of multiple simultaneous equations. A simple approach for a solution method would require solving explicitly for one variable in terms of the other variables and using that expression in the other equations. This process would be repeated for each of the other variables defined.

Looking at the matrix A, it is not obvious upon inspection that the matrix is full rank. The determinant of this matrix is also a value of 1, so the inverse of this matrix does exist. The inverse matrix is:

$$A^{-1} = \begin{bmatrix} -2 & -1 & -1 & -1 & -1 \\ 2 & 1 & 1 & 2 & 2 \\ 1 & 1 & 1 & 1 & 1 \\ -1 & -1 & 0 & -1 & -1 \\ -2 & -1 & -1 & -2 & -1 \end{bmatrix}$$

It can be verified that this is truly the matrix inverse by multiplying it with matrix A. The result should be a 5x5 identity matrix, since:

$$AA^{-1} = I$$

7.8.1   MATLAB Code Example

```
%
% Consider a reactor system with three species
% The species are just labled a, b, and c
% The reactor system has two reactors.
% Each reactor has different reactions taking
% place. Two reactions take place in reactor 1
% and only one reaction takes place in reactor 2.
%
% Define variables
% na0, nb0, nc0 molar feed rates, mol/hr
% na1, nb1, nc1 rates exiting reactor 1, mol/hr
% na2, nb2, nc2 rates exiting reactor 2, mol/hr
% z1 extent of reaction 1, a to b and c, mol/hr
% z2 extent of reaction 2, b to c, mol/hr
% z3 extent of reaction 3, a to b, mol/hr
% Reaction 1 and 2 take place in reactor 1.
% Reaction 3 takes place in reactor 2
%

% Start with empty matrices
A=[];b=[];

% Adding a new row to A and b adds a new equation!
% This is used by augmenting the matrix or vector with
% a new row using this syntax: A=[A ; [ 1 0 1] ]

% specify inlet mole rates! na0 = 100, nb0= 20, nc0 = 5
A=[A ; [ 1 0 0 0 0 0 0 0 0 0 0 0]];
b=[b ; 100];
A=[A ; [ 0 1 0 0 0 0 0 0 0 0 0 0]];
b=[b ; 20];
A=[A ; [ 0 0 1 0 0 0 0 0 0 0 0 0]];
b=[b ; 5];

% specify inlet mole rates! na2 = 50, nb2= 55, nc2 = 50
A=[A ; [ 0 0 0 0 0 0 1 0 0 0 0 0]];
b=[b ; 50];
A=[A ; [ 0 0 0 0 0 0 0 1 0 0 0 0]];
b=[b ; 55];
A=[A ; [ 0 0 0 0 0 0 0 0 1 0 0 0]];
b=[b ; 50];

% specify balances on first reactor
A=[A ; [ 1 0 0 -1 0 0 0 0 0 -1 0 0]];
b=[b ; 0];
```

```
A=[A ; [ 0 1 0 0 -1 0 0 0 0 1 -1 0]];
b=[b ; 0];
A=[A ; [ 0 0 1 0 0 -1 0 0 0 1 1 0]];
b=[b ; 0];

% specify balances on second reactor
A=[A ; [ 0 0 0 1 0 0 -1 0 0 0 0 -1]];
b=[b ; 0];
A=[A ; [ 0 0 0 0 1 0 0 -1 0 0 0 1]];
b=[b ; 0];
A=[A ; [ 0 0 0 0 0 1 0 0 -1 0 0 0]];
b=[b ; 0];

% specify overall balances
A=[A ; [ 1 0 0 0 0 0 -1 0 0 -1 0 -1]];
b=[b ; 0];
A=[A ; [ 0 1 0 0 0 0 0 -1 0 1 -1 1]];
b=[b ; 0];
A=[A ; [ 0 0 1 0 0 0 0 0 -1 1 1 0]];
b=[b ; 0];

% Does it solve? Check without overall balances?
AA=A(1:12,:);bb=b(1:12);
inv(AA)*bb

%Show overall balance is linear combo of R1 and R2
A(7:9,:)+A(10:12,:)-A(13:15,:)

% Check to see if you can drop an equation and still keep
  full rank 12

for i=1:12
  r=rank([A(1:i-1,:) ; A(i+1:end,:) ]);
  if r<12
    disp(['Equation ' num2str(i) ' required!'])
  end
end

% The first six equations are required! This makes
% sense as they are simple equations that are explicit in
% one variable. These equations are how the variables are defined.

% Can you pick any three equations to remove? No!
% Can you drop all balances for c? Drop equations 9, 12, 15?
keep=[ 1 2 3 4 5 6 7 8 10 11 13 14];
rank(A(keep,:))
% This is only rank 10, which means you can't solve for all
% of the 12 variables
```

```
% Dropping overall balance works though!
keep=[ 1 2 3 4 5 6 7 8 9 10 11 12];
rank(A(keep,:))

% Drop A bal in r1, B bal in r2, C bal overall
keep=[ 1 2 3 4 5 6 8 9 10 12 13 14 ];
rank(A(keep,:))

% The key section includes the balances on r1 and r2,
% assuming feed and product concentrations are fixed:
AA=A(7:12,[4 5 6 10 11 12])
% The first six rows can be removed because they are only
% used to specify the inlet and outlet flows. Similarly,
% columns corresponding to variables for inlet and outlet
% flows can be removed.


%
% In the resulting set of equations, almost every
% row involves two variables! Nothing is explicit,
% except for the flow of C between reactors since C does
% not react in the second reactor
%
% If we know the values for na0, nb0, nc0 and na2,
% nb2, and nc2, we can substitute them into the
% equations and find the values for the RHS
bb=[-100 -20 -5 50 55 50]'

% Check that our set of six equations are valid:
rank(AA) % Six eqns with six vars must be rank six
det(AA) % determinant of the matrix must be nonzero

% And now you can solve these six equations:
inv(AA)*bb

% This matches the result from the previous with
% twelve variables. Removing equation 6 and variable
% nc1 results in a 5x5 set of equations:
AAA=A(7:11,[4 5 10 11 12 ]);
bbb=[-100 -20 45 50 55 ]';
inv(AAA)*bbb

rank(AAA)
det(AAA)

% Matrix inverse example
%
% The matrix inverse can be found using row reduction
% techniques. To find the inverse of matrix A, first
% set up a set of equations in the form:
```

```
%        A | I
% Where I represents the identity matrix.
%
% Row reduction can be used to find the inverse.
% Row reduction operations include the following:
% 1. Swap any two rows
% 2. Multiply any row by a constant
% 3. Add or subtract any row to another
% Rules can be used simultaneously in a single step.
%
% From the form A|I row reduction operations are used
% to manipulate the set of equations until the values
% are in the form: I|Ainv
% To do this, consider the first column initially,
% getting a 1 on the diagonal, then get a value of 0
% below the diagonal element. Then move to column 2,
% get a 1 on the diagonal, then get values of 0 below.
% Continue on until you only have values of 1 on the
% diagonal and 0s below. Then start in the last column
% and use the last row to get 0s above the
% diagonal. This part may require two operations at
once, like
% subtracting five times row 5 from row 4. After the
% last column has only 0 above the diagonal, move
% left to the second to last column and use the second
to
% last row to get 0s above the diagonal. Continue
until there
% are only values of 1 on the diagonal.

% Set up the matrix with the original matrix next to
% an identity matrix of the same size:
M=[AAA eye(5)]

% Rearrange rows 4 and 5 to 1 and 2, move R4 to R1
and R5 to R2
M=[M(4:5,:) ; M(1:3,:)]

% Work on clearing out column 1 below the diagonal
% Add R1 to R3
M(3,:)=M(3,:)+M(1,:)

% Work on clearing out column 2 below the diagonal
% Add R2 to R4
M(4,:)=M(4,:)+M(2,:)

% Work on clearing out column 3 below the diagonal
```

```
% Mult R3 by -1
M(3,:)=M(3,:)*-1

% Subtract R3 from R4
M(4,:)=M(4,:)-M(3,:)

% Subtract R3 from R5
M(5,:)=M(5,:)-M(3,:)

% Work on clearing out column 4 below the diagonal
% Mult R4 by -1
M(4,:)=M(4,:)*-1

% Subtract R4 from R5
M(5,:)=M(5,:)-M(4,:)

% Work on getting a 1 on the diagonal in the 5th row
% Mult R5 by -1
M(5,:)=M(5,:)*-1

% Now, move to column 5 and remove values above the
diagonal
% Subtract R5 from R3
M(3,:)=M(3,:)-M(5,:)

% Subtract R5 from R2
M(2,:)=M(2,:)-M(5,:)

% Add R5 to R1
M(1,:)=M(1,:)+M(5,:)

% If it were needed, you would move to column 4 and
use
% row 4 to clear out the values above the diagonal
% in column 4. Then move on to column 3 and then 2.

% Extract the inverse of matrix AAA from the matrix
of coefficients
AAAinv=M(:,6:end)
inv(AAA) % does it match what MATLAB calculates?
inv(AAA)-AAAinv % If it matches, this should be all
0s.
```

Problems

7.1 Given the vectors a and b, determine the dot product $a \cdot b$

$$a = \begin{bmatrix} 1 \\ 2 \\ 7 \end{bmatrix} \quad b = \begin{bmatrix} -3 \\ 2 \\ 1 \end{bmatrix}$$

7.2 Given the vectors c and d, determine the dot product $c \cdot d$

$$c = \begin{bmatrix} 0 \\ 1 \\ -1 \end{bmatrix} \quad d = \begin{bmatrix} 5 \\ 3 \\ 4 \end{bmatrix}$$

7.3 Given the vectors f and g, determine the dot product $f \cdot g$

$$f = \begin{bmatrix} 1 \\ 2 \\ -1 \end{bmatrix} \quad g = \begin{bmatrix} 4 \\ 2 \\ 8 \end{bmatrix}$$

7.4 Given the vectors x and y, determine the dot product $x \cdot y$

$$x = \begin{bmatrix} -1 \\ -2 \\ -3 \end{bmatrix} \quad y = \begin{bmatrix} 2 \\ -1 \\ 1 \end{bmatrix}$$

7.5 Given the vectors x_1 and x_2, determine the dot product $x_1 \cdot x_2$

$$x_1 = \begin{bmatrix} 3 \\ 2 \\ 1 \end{bmatrix} \quad x_2 = \begin{bmatrix} 1 \\ 0 \\ -3 \end{bmatrix}$$

7.6 Given the vectors a_1 and a_2, are the vectors orthogonal?

$$a_1 = \begin{bmatrix} 1 \\ 2 \\ 1 \end{bmatrix} \quad a_2 = \begin{bmatrix} 5 \\ 0 \\ 4 \end{bmatrix}$$

7.7 Given the vectors b and c, are the vectors orthogonal?

$$b = \begin{bmatrix} 3 \\ 7 \\ -5 \end{bmatrix} \quad c = \begin{bmatrix} -1 \\ -1 \\ -2 \end{bmatrix}$$

7.8 Calculate the determinant of the following matrix:

$$A = \begin{bmatrix} 1 & 4 \\ -2 & -3 \end{bmatrix}$$

7.9 Calculate the determinant of the following matrix:

$$B = \begin{bmatrix} -2 & -3 \\ 6 & 1 \end{bmatrix}$$

7.10 Calculate the determinant of the following matrix:

$$C = \begin{bmatrix} 0 & 2 \\ -7 & 1 \end{bmatrix}$$

7.11 Calculate the determinant of the following matrix:

$$D = \begin{bmatrix} 1 & -2 & 0 \\ 1 & 5 & 2 \\ 7 & 1 & 3 \end{bmatrix}$$

7.12 Calculate the determinant of the following matrix:

$$E = \begin{bmatrix} 0 & -1 & 2 \\ 1 & 1 & 3 \\ 0 & 1 & 1 \end{bmatrix}$$

7.13 Given the vectors a and b, determine the cross product $a \times b$

$$a = \begin{bmatrix} 1 \\ 2 \\ 7 \end{bmatrix} \quad b = \begin{bmatrix} -3 \\ 2 \\ 1 \end{bmatrix}$$

7.14 Given the vectors c and d, determine the cross product $c \times d$

$$c = \begin{bmatrix} 0 \\ 1 \\ -1 \end{bmatrix} \quad d = \begin{bmatrix} 5 \\ 3 \\ 4 \end{bmatrix}$$

7.15 Given the vectors f and g, determine the cross product $f \times g$

$$f = \begin{bmatrix} 1 \\ 2 \\ -1 \end{bmatrix} \quad g = \begin{bmatrix} 4 \\ 2 \\ 8 \end{bmatrix}$$

7.16 Given the vectors x and y, determine the cross product $x \times y$

$$x = \begin{bmatrix} -1 \\ -2 \\ -3 \end{bmatrix} \quad y = \begin{bmatrix} 2 \\ -1 \\ 1 \end{bmatrix}$$

7.17 Given the vectors x_1 and x_2, determine the cross product $x_1 \times x_2$

$$x_1 = \begin{bmatrix} 3 \\ 2 \\ 1 \end{bmatrix} \quad x_2 = \begin{bmatrix} 1 \\ 0 \\ -3 \end{bmatrix}$$

7.18 Calculate the following values:

$$\begin{bmatrix} 1 & -2 & 0 \\ -1 & 5 & -2 \\ 7 & -1 & 3 \end{bmatrix} \begin{bmatrix} 1 \\ 4 \\ 3 \end{bmatrix}$$

7.19 Calculate the following values:

$$\begin{bmatrix} 0 & 2 \\ -7 & 1 \end{bmatrix} \begin{bmatrix} 1 & 0 & 2 & -1 \\ 0 & -1 & 1 & 3 \end{bmatrix}$$

7.20 Calculate the following values:

$$\begin{bmatrix} -1 & -3 & 0 \\ 0 & 2 & 4 \end{bmatrix} \begin{bmatrix} -1 \\ 2 \\ 1 \end{bmatrix}$$

7.21 Calculate the following values:

$$\begin{bmatrix} 1 & -2 \\ -2 & 4 \end{bmatrix} \begin{bmatrix} 1 & 2 \\ 3 & 2 \end{bmatrix}$$

7.22 Calculate the following values:

$$\begin{bmatrix} 0 & -2 & 0 \\ -1 & 0 & 0 \\ 3 & -3 & 0 \end{bmatrix} \begin{bmatrix} -1 \\ -2 \\ -9 \end{bmatrix}$$

7.23 Calculate the following values:

$$\begin{bmatrix} 0 & 1 \\ -2 & 3 \end{bmatrix} \begin{bmatrix} 1 & 0 & 4 & -1 \\ 1 & -1 & 2 & -3 \end{bmatrix}$$

7.24 Solve the following set of equations using row reduction methods.

$$x + y + z = 2$$
$$2x + 4y + 6z = 12$$
$$-2y + 6z = 2$$

7.25 Solve the following set of equations using row reduction methods.

$$x + z = 1$$
$$8y + 3z = 1$$
$$2y + 3z = 1$$

7.26 Solve the following set of equations using row reduction methods.

$$-x + 2y + z = 7$$
$$-3x + 2y - z = 1$$
$$2x + y + z = 7$$

7.27 Find the inverse of the following matrix:

$$\begin{bmatrix} 2 & 1 \\ -2 & 4 \end{bmatrix}$$

7.28 Find the inverse of the following matrix:

$$\begin{bmatrix} 1 & 1 \\ 3 & 2 \end{bmatrix}$$

7.29 Find the inverse of the following matrix:

$$\begin{bmatrix} 1 & 0 & 1 \\ 0 & 8 & 3 \\ 0 & 2 & 3 \end{bmatrix}$$

7.30 Find the inverse of the following matrix:

$$\begin{bmatrix} 1 & 2 & 1 \\ 2 & 4 & 3 \\ 0 & 2 & 3 \end{bmatrix}$$

Further Reading

1. Axler, S. (2015). *Linear algebra done right* (3rd ed.). Springer.
2. Larson, R. (2016). *Elementary linear algebra* (8th ed.). Cengage Learning.
3. Lay, D.C., Lay, S.R., & McDonald, J.J. (2014). *Linear algebra and its applications* (5th ed.). Pearson.
4. Savov, I. (2017). *No bullshit guide to linear algebra* (2nd ed.). Minireference Co.
5. Singh, K. (2013). *Linear algebra: Step by step* (1st ed.). Oxford University Press.
6. Strang, G. (1993). *Introduction to linear algebra*. Wellesley, MA: Wellesley-Cambridge Press.

Chapter 8
Root Finding and Integration

8.1 Finding Roots of an Equation

There are many reasons to find a root of an equation. The following are some examples of engineering calculations which can be aided by numerical solution of $f(x) = 0$.

8.1.1 Real Gas

Consider the ideal gas law $PV = nRT$. A more accurate version to determine the pressure of a real system is given by:

$$P = \frac{RT}{\frac{V}{n} - b} - \frac{a}{\sqrt{T}\,\frac{V}{n}\left(\frac{V}{n} + b\right)}$$

where a and b are constants that depend on the type of gas. Given a known volume V, a known number of moles n, and values for a and b, the pressure P becomes a function of temperature:

$$P = f(T)$$

Given a target pressure, the problem is to determine the temperature value that makes the gas attain a specified pressure.

© Springer Nature Switzerland AG 2022
E. Gatzke, *Introduction to Modeling and Numerical Methods for Biomedical and Chemical Engineers*, https://doi.org/10.1007/978-3-030-76449-4_8

8.1.2   Falling Droplet

The velocity of a falling droplet may be derived as:

$$v = \sqrt{\frac{4gd}{3C_d}\left(\frac{\rho_s - \rho}{\rho}\right)}$$

where g is the gravitational acceleration constant, d is the droplet size, ρ_s is the droplet density, ρ is the gas density, and C_d is the drag coefficient. Assuming the droplet size and density are known along with the drag coefficient, the velocity of the droplet is a function of the gas density ρ:

$$v = f(\rho)$$

Given a target velocity, the problem becomes to find the density value that correlates to the desired velocity.

8.1.3   General Form

Many problems can be stated in equation form where one variable is a function of another. Often, the dependent variable is y and the independent variable is x. The relationship between the two is written

$$y = f(x)$$

A root of this function is found when $y = 0$. In general, the problem is defined as searching for values of x such that

$$f(x) = 0$$

In some cases, the problem could be stated as searching for the point where two functions g and h are equal:

$$g(x) = h(x)$$

This can be easily transformed to the form $f(x) = 0$ as:

$$g(x) - h(x) = 0$$

where $f(x)$ is now defined as:

$$f(x) = g(x) - h(x)$$

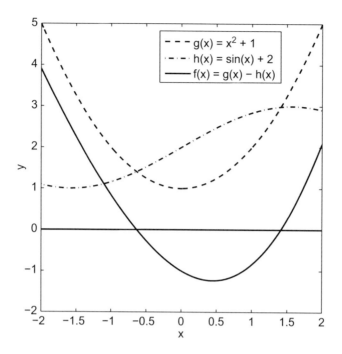

Fig. 8.1 Demonstration of conversion of $g(x) = h(x)$ to the form $f(x) = g(x) - h(x)$. Note that where $f(x)$ crosses 0 correlates to x values where $g(x) = h(x)$

This transformation is similar to finding the value of x where the function is equal to a constant c instead of 0. In this case, the problem becomes:

$$g(x) = c$$

which then becomes:

$$f(x) = g(x) - c$$

which can then be used as $f(x)$ in a root finding method (Fig. 8.1).

8.2 Newton's Method

Newton's method requires that you know the following:

- A scalar function of one variable: $y = f(x)$
- The derivative of the function: $f'(x) = \frac{df}{dx}(x)$
- An initial value for x

The next value for x can be found using the following equation:

$$x_{n+1} = x_n - \frac{f(x_n)}{f'(x_n)}$$

This may be established using a Taylor series expansion of the nonlinear function at the value x_n or using arguments related to right triangles. The function evaluated at x_n is $f(x_n)$. This represents how far above or below 0 the function is at point x_n. The slope of that function at x_n is given by $\frac{df}{dx}(x_n)$ and is sometimes written as $f'(x_n)$. Given this information, a line tangent to the original function can be constructed as:

$$y = \frac{df}{dx}(x_n)(x_{n+1} - x_n) + f(x_n)$$

However, we are looking to find roots of the equation $y = f(x)$ or the value of x where $f(x) = 0$. Using the tangent function line, one may set y to 0 and determine the next value of x, x_{n+1}:

$$0 = \frac{df}{dx}(x_n)(x_{n+1} - x_n) + f(x_n)$$

$$-f(x_n) = \frac{df}{dx}(x_n)(x_{n+1} - x_n)$$

$$-\frac{f(x_n)}{\frac{df}{dx}(x_n)} = x_{n+1} - x_n$$

$$x_n - \frac{f(x_n)}{\frac{df}{dx}(x_n)} = x_{n+1}$$

Resulting in the general form equation:

$$x_{n+1} = x_n - \frac{f(x_n)}{f'(x_n)}$$

Newton's method converges quite rapidly. This means that $f(x) \approx 0$ after only a few iterations in most cases. If there are multiple roots to an equation, Newton's method may converge to any of the real roots, depending on the initial starting value for x_n.

There are some cases where Newton's method will not converge. If the derivative of the function goes to zero at any point, Newton's method would fail as:

$$x_{n+1} = x_n - \frac{f(x_n)}{0} = x_n - \infty$$

This makes sense. If the tangent line slope goes to zero when $f(x_n) \neq 0$, then the tangent line is parallel to the line $y = 0$. The two lines are parallel and never intersect! Similarly, some functions may diverge or oscillate, never converging to a root if the initial guess is not adequately close to a solution (Figs. 8.2 and 8.3).

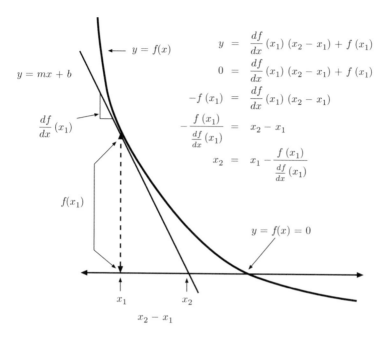

Fig. 8.2 The first iteration in solving $y = f(x) = 0$ using Newton's method

Example 8.1 (Newton's Method) Consider the function $f(x) = x^3 + x + 1$ starting from the value $x_1 = 2$. The derivative of this function is given by $f'(x) = 3x^2 + 1$. The iterative update equation is:

$$x_{n+1} = x_n - \frac{f(x_n)}{f'(x_n)}$$

so that the next value of x can be found just by plugging in the values of x_1, $f(x_1)$, and $f'(x_1)$. At $x_1 = 2$:

$$f(2) = 2^3 + 2 + 1 = 11$$
$$f'(2) = 3(2)^2 + 1 = 13$$

and the new value, x_2 is found as:

$$x_2 = 2 - \frac{11}{13} = 1.154$$

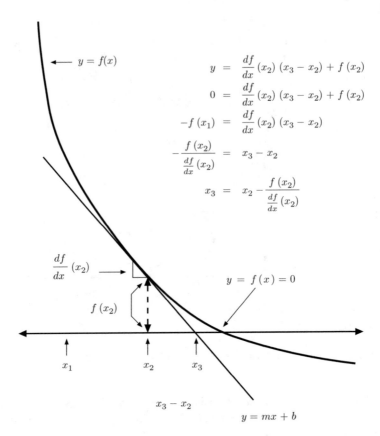

$$y = \frac{df}{dx}(x_2)\,(x_3 - x_2) + f(x_2)$$

$$0 = \frac{df}{dx}(x_2)\,(x_3 - x_2) + f(x_2)$$

$$-f(x_1) = \frac{df}{dx}(x_2)\,(x_3 - x_2)$$

$$-\frac{f(x_2)}{\frac{df}{dx}(x_2)} = x_3 - x_2$$

$$x_3 = x_2 - \frac{f(x_2)}{\frac{df}{dx}(x_2)}$$

Fig. 8.3 The second iteration in solving $y = f(x) = 0$ using Newton's method

iteration n	x_n	$f(x_n)$	$f'(x_n)$	x_{n+1}
1	2	11	13	1.154
2	1.154	3.690	4.994	0.415
3	0.415	1.486	1.517	−0.565
4	−0.665	0.254	1.958	−0.695
5	−0.695	−0.0308	2.449	−0.683
6	−0.683	−0.0003	2.397	

This problem gets fairly close after five iterations. The value for x_6 is -0.683. Evaluating the function at this value $f(-0.683) = -0.0003$. Additional iterations could produce more accurate results.

Newton's method can quickly be implemented in a variety of ways. Since the number of required iterations is not known, the method should run until $f(x)$ is close to 0. A simple way to check this condition is to look at the magnitude of the absolute value of $f(x)$ at each iteration. In pseudo code, this may appear as:

x_{new} =Initial value for x
WHILE $|f(x_{new})| > \epsilon$
$\quad x_{old} = x_{new}$
$\quad x_{new} = x_{old} - \frac{f(x_{old})}{f'(x_{old})}$
END

Converting this to simple MATLAB code is straightforward. Using anonymous functions to define the function and the derivative function leads to:

```
f=@(x) x^3+x+1
fprime=@(x) 3*x^2+1
xnew=2
while (abs(f(xnew))>.000001)
    xold=xnew;
    xnew=xold-f(xold)/fprime(xold);
end
disp([' Root at x = ' num2str(xnew) ])
disp([' f(x)= ' num2str(f(xnew)) ])
```

The biggest drawback when using Newton's method is the requirement to have an expression for the derivative function. However, the derivative can be approximated by evaluating the function at point close to x_n. For a small value of ϵ, the derivative is:

$$f'(x_n) \approx \frac{\Delta f}{\Delta x} = \frac{f(x_n + \epsilon) - f(x_n)}{(x_n + \epsilon) - x_n} = \frac{f(x_n + \epsilon) - f(x_n)}{\epsilon}$$

Without actually explicitly determining the value of the derivative function, the previous MATLAB code becomes:

```
f=@(x) x^3+x+1
epsilon=0.00001
xnew=2
while (abs(f(xnew))>.000001)
    xold=xnew;
    fprime=(f(xold+epsilon)-f(xold))/epsilon;
    xnew=xold-f(xold)/fprime;
end
disp([' Root at x = ' num2str(xnew) ])
disp([' f(x)= ' num2str(f(xnew)) ])
```

8.3 Bisection Method

Newton's method will rapidly converge to a solution in most cases. However, Newton's method is not guaranteed to converge in all cases. The bisection method may be more robust. Bisection is also easy to implement, as the function derivative is not needed. Assuming the following conditions:

- The function $f(x)$ is continuous over the range of interest
- The function is known to have a root on $[x_L \leq x \leq x_U]$

One easy way to check if there is a root between $x = x_L$ and $x = x_R$ is to use the product of the function evaluated at both end points. If the product $f(x_L) f(x_U)$ is positive, then $f(x_L))$ and $f(x_U)$ are either both positive or both negative. If the product $f(x_L) f(x_U)$ is negative, then $f(x_L)$ and $f(x_U)$ are have different signs and a root must be between x_L and x_U if the function is continuous.

Assuming that the function is continuous and there is a root between $x = x_L$ and $x = x_R$, the values for x_L and x_U are updated as follows:

$x_M = x_L + \frac{x_U - x_L}{2}$
IF $f(x_L) f(x_M) > 0$
$\quad x_L = x_M$
ELSE
$\quad x_U = x_M$
END

The bisection method is also easily converted to simple MATLAB code. Using the inline function to define the function leads to:

```
f=@(x) x^3+x+1
xl=-2
xu=2
xm=xl+(xu-xl)/2
while (abs(f(xm))>0.0001)
        disp([xl f(xl) xu f(xu)]);
        xm=xl+(xu-xl)/2;
        if ( f(xl)*f(xm) >0 )
            xl=xm;
        else
            xu=xm;
        end
    end
```

Note that the bisection code does not require knowledge of the derivative function. In many cases, determining the derivative function can be difficult or impossible. For example, some engineering problems require solution of a complex simulation. In these cases, the function evaluation is given by providing a value x, then running the complex simulation to determine the resulting value for $f(x)$. No derivative can easily be found. In these situations, derivative-free methods make sense.

Example 8.2 (Bisection M Method) Consider the function $f(x) = x^3 + x + 1$ again starting with the interval $x_L = -2$ to $x_U = 2$. Evaluating the function at the endpoints $f(-2) = -9$ and $f(2) = 11$. Since the function evaluations are negative and positive, there should be at least one root between $x = -2$ and $x = 2$. The midpoint can be found as $x_L + \frac{x_U - x_L}{2}$. The midpoint is obviously at $x_M = 0$ where $f(0) = 1$. Using the bisection method, it takes more iterations to converge to a solution.

iteration	x_L	$f(x_L)$	x_U	$f(x_U)$
1	-2	-9	2	11
2	-2	-9	0	1
3	-1	-1	0	1
4	-1	-1	-0.5	0.375
5	-0.75	-0.1719	-0.5	0.375
6	-0.75	-0.1719	-0.625	0.1309
7	-0.6875	-0.0125	-0.625	0.1309
8	-0.6875	-0.0125	-0.6563	0.0611
9	-0.6875	-0.0125	-0.6719	0.0248
10	-0.6875	-0.0125	-0.6797	0.0063

8.4 Multivariate Root Finding

In some cases, you may have multiple relationships involving multiple variables. In general, if you have N variables, you need N equations to determine a solution. Recall Newton's method for a single scalar function:

$$x_{n+1} = x_n - \frac{f(x_n)}{f'(x_n)}$$

This equation appears similar to the multivariate version of Newton's method:

$$\underline{x}_{n+1} = \underline{x}_n - \left(J_F\left(\underline{x}_n\right)\right)^{-1} F\left(\underline{x}_n\right)$$

Here, the vector of N unknown variables at iteration n is $\underline{x}_n$ and $F(\underline{x})$ is the N-dimensional vector of functions defined as:

$$F(\underline{x}) = \begin{bmatrix} f_1(\underline{x}) \\ f_2(\underline{x}) \\ \vdots \\ f_N(\underline{x}) \end{bmatrix}$$

where we desire to find the vector $\underline{x}$ where $F(\underline{x}) = 0$. Multivariate Newton's method also requires that the Jacobian matrix J_F be determined. This matrix represents all the partial derivatives for the N functions with respect to all N variables:

$$J_F(\underline{x}) = \begin{bmatrix} \frac{\partial f_1}{\partial x_1} & \frac{\partial f_1}{\partial x_2} & \cdots & \frac{\partial f_1}{\partial x_N} \\ \frac{\partial f_2}{\partial x_1} & \frac{\partial f_2}{\partial x_2} & & \frac{\partial f_2}{\partial x_N} \\ \vdots & & \ddots & \vdots \\ \frac{\partial f_N}{\partial x_1} & \frac{\partial f_N}{\partial x_2} & \cdots & \frac{\partial f_N}{\partial x_N} \end{bmatrix}$$

At each iteration, the vector multivariate function $F(\underline{x})$ is evaluated at $\underline{x}_n$, the Jacobian matrix $J_F(\underline{x})$ is evaluated at $\underline{x}_n$, and the inverse of the Jacobian matrix is determined.

Just as in the scalar version of Newton's method, the nonlinear function $F(\underline{x})$ is being approximated in a linear form that is easier to solve. We desire to find a point $\underline{x}$ where:

$$F(\underline{x}) = 0$$

The multivariate function can be approximated at the current point $\underline{x}_n$ using the multivariate first-order Taylor series as:

$$F(\underline{x}_{n+1}) = F(\underline{x}_n) + J_F(\underline{x}_n)(\underline{x}_n - \underline{x}_{n+1})$$

Setting $F(\underline{x}_{n+1}) = 0$, this becomes a linear algebra problem in the form $\underline{\underline{A}}\,\underline{x} = \underline{b}$ where you are solving at iteration n for the new value $\underline{x}_{n+1}$ as:

$$F(\underline{x}_{n+1}) = F(\underline{x}_n) + J_F(\underline{x}_n)(\underline{x}_{n+1} - \underline{x}_n)$$
$$0 = F(\underline{x}_n) + J_F(\underline{x}_n)(\underline{x}_{n+1} - \underline{x}_n)$$
$$-F(\underline{x}_n) = J_F(\underline{x}_n)(\underline{x}_{n+1} - \underline{x}_n)$$
$$-(J_F(\underline{x}_n))^{-1}F(\underline{x}_n) = (J_F(\underline{x}_n))^{-1}J_F(\underline{x}_n)(\underline{x}_{n+1} - \underline{x}_n)$$
$$-(J_F(\underline{x}_n))^{-1}F(\underline{x}_n) = (\underline{x}_{n+1} - \underline{x}_n)$$
$$\underline{x}_{n+1} = \underline{x}_n - (J_F(\underline{x}_n))^{-1}F(\underline{x}_n)$$

Example 8.3 (Multivariate Solution of F(x) = 0) Consider a simple problem looking for the solution of the following equations:

$$f_1(x, y) = x^2 + y^2 - 9$$
$$f_2(x, y) = y - x^2$$

These equations represent looking for a point where the parabola intersects a circle around the origin with radius 3.

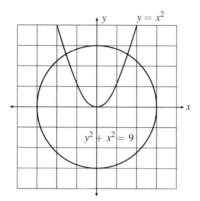

For this problem, the unknown vector $\underline{x}_n$ is $[x \ y]^T$ and the vector function $F\left(\underline{x}_n\right)$ is:

$$F\left(\underline{x}_n\right) = \begin{bmatrix} x^2 + y^2 - 9 \\ y - x^2 \end{bmatrix}$$

The Jacobian matrix for function $F\left(\underline{x}_n\right)$ is given by:

$$J_F\left(\underline{x}_n\right) = \begin{bmatrix} 2x & 2y \\ -2x & 1 \end{bmatrix}$$

Starting at the point $x = 2$, $y = 1$ yields the following table of the first three iterations. One final solution is at $x = 1.594$, $y = 2.541$.

n	$\underline{x}_n$	$F\left(\underline{x}_n\right)$	$J_F\left(\underline{x}_n\right)$
0	$\begin{bmatrix} 2 \\ 1 \end{bmatrix}$	$\begin{bmatrix} -4 \\ -4 \end{bmatrix}$	$\begin{bmatrix} 4 & 2 \\ -4 & 1 \end{bmatrix}$
1	$\begin{bmatrix} 1.833 \\ 3.333 \end{bmatrix}$	$\begin{bmatrix} 5.472 \\ -0.028 \end{bmatrix}$	$\begin{bmatrix} 3.667 & 6.667 \\ -3.667 & 1.000 \end{bmatrix}$
2	$\begin{bmatrix} 1.632 \\ 2.623 \end{bmatrix}$	$\begin{bmatrix} 0.545 \\ -0.041 \end{bmatrix}$	$\begin{bmatrix} 3.264 & 5.246 \\ -3.264 & 1.000 \end{bmatrix}$

The first two iterations are showing here graphically, including the linear approximation derived at the first two steps:

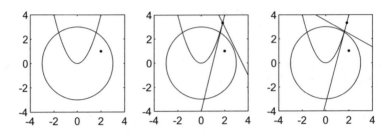

Note that the initial point $x = 2$, $y = 1$ does not satisfy either function, f_1 or f_2. Also, note that $F\left(\underline{x_n}\right)$ is getting closer to $(0, 0)$ at each iteration as the method converges.

When solving a scalar problem, you know the function $f(x)$ which generates a value for y at a given value of x. A one-dimensional line in the form $y = mx + b$ can be used to find the next iteration by setting $y = 0$ and solving for the new x. In the multivariate case, the vector function $F\left(\underline{x_n}\right)$ will generate N values for a single N-dimensional point $\underline{x_n}$. In the two-dimensional problem, each of the two functions can be thought of as returning a value in the z plane for values of x and y. For the previous example, the parabolas and circles shown represent the only points in all of x, y that satisfy $z = 0$. When the linear approximation is applied to either the parabola or the circle function in the two-dimensional problem, the equation of a 2D plane tangent to the individual multivariate function is determined. The straight lines shown in the example represent the intersection of the 2D flat planes with the plane $z = 0$.

Consider the function $z = y - x^2$ which represents a nonlinear two-dimensional surface in 3D space. Taking the first-order Taylor series at a point such as $x = 2$, $y = 1$ generates a 2D flat plane in 3D space defined for values of x and y. One row in the $\underline{\underline{A}}\,\underline{x} = \underline{b}$ approximation is equivalent to:

$$f(x, y) \approx f(x, y)|_{x=x^*, y=y^*} + \frac{\partial f}{\partial x}(x, y)|_{x=x^*, y=y^*}\left(x - x^*\right)$$
$$+ \frac{\partial f}{\partial y}(x, y)|_{x=x^*, y=y^*}\left(y - y^*\right)$$

For the function $z = y - x^2$, evaluated at $x^* = 2$, $y^* = 1$ the definition of the 2D plane in 3D space is given by:

$$z = \left(1 - 2^2\right) + (-2(2))(x - 2) + (1)(y - 1)$$

The nonlinear surface, the tangent plane, and the plane for $z = 0$ are shown in Fig. 8.4.

Fig. 8.4 The nonlinear surface $z = y - x^2$, the tangent plane at $x = 2$, $y = 1$, and the plane for $z = 0$

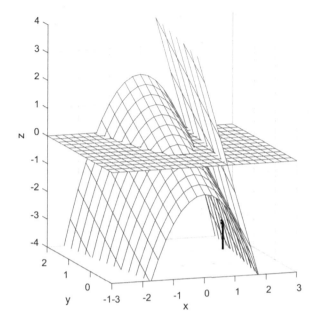

8.5 Integrating a Function

There are many cases where determination of the integral of a function is important. The integral can also be considered the Area Under a Curve (AUC). In a calculus class, various analytical methods are considered for determining the integral values of various functions. However, in many engineering cases, the functions are very complex or have no known close-form solution for the integral. This means that numerical methods must be derived.

8.5.1 Average Velocity

Given the average velocity for a liquid flowing vertically through a pipe:

$$v_y = f(x)$$

The average value can be found by integrating the velocity in the x direction from $x = 0$ to $x = L$ and dividing by the total width of the pipe.

Note that velocity can be measured in $\frac{m}{s}$. The integral in the x direction would have units of $\frac{m^2}{s}$ but dividing by the total length gives units of velocity, $\frac{m}{s}$, the average value for v_y.

8.5.2　Average Concentration

Given that the concentration at each level in a reactor is a function of the tank height:

$$c = f(h)$$

The average concentration can be found by integrating the function with respect to height from $h = 0$ to $h = H$ and dividing by the total tank height.

Note that concentration is measured in moles per volume, or simply $\frac{mol}{m^3}$. The integral of concentration along one direction would have units of $\frac{mol}{m^2}$. Dividing the integral value by the total length gives units of average concentration, $\frac{mol}{m^3}$.

Example 8.4 (Average Concentration) The average concentration in a reactor is unknown. It is assumed that the concentration in a reactor follows the following model as a function of tank height:

$$c(h)\left(\frac{mol}{L}\right) = \left(0.03 + 0.001h^2\right)\left(\frac{mol}{L}\right)$$

Where concentration is in $\frac{mol}{L}$ and height is in m. First, we must convert the units. There are $1000\,L$ in 1 cubic meter. The concentration in units of $\frac{mol}{m^3}$ should be much larger, as you have many more moles in a cubic meter of material.

$$c(h)\left(\frac{mol}{m^3}\right) = \left(0.03 + 0.001h^2\right)\left(\frac{mol}{L}\right)\left(\frac{1000\,L}{1\,m^3}\right)$$

The concentration as a function of height is now:

$$c(h)\left(\frac{mol}{m^3}\right) = \left(30 + 1h^2\right)\left(\frac{mol}{m^3}\right)$$

This also implies that the model parameter 30 and 1 have units so that each term is in units of $\frac{mol}{m^3}$. In this case, the model could be written as:

$$c(h)\left(\frac{mol}{m^3}\right) = \left(30\left(\frac{mol}{m^3}\right) + 1\left(\frac{mol}{m^5}\right)(h(m))^2\right)$$

Given that the reactor height is 6 m, the integral becomes:

$$\int_{h=0}^{h=4} c(h)\,dh = 30h + \frac{1}{3}h^3\Big|_{h=0}^{h=6}$$

$$= 120 + \frac{216}{3}\frac{mol}{m^2} = 192\,\frac{mol}{m^2}$$

To find the average concentration, one must divide by the reactor height.

$$\bar{c} = \frac{192 \frac{\text{mol}}{\text{m}^2}}{6\,\text{m}} = 192\,\frac{\text{mol}}{\text{m}^2}\,\frac{1}{6\text{m}} = 32\,\frac{\text{mol}}{\text{m}^3} = 0.032\,\frac{\text{mol}}{\text{L}}$$

8.5.3   Total Drug Delivered

Consider the continuous infusion insulin pump. Diabetic patients can use this device to receive a continuous infusion of insulin to help moderate their blood glucose levels. Given that the flow rate for the pump is measured as a function of time:

$$F = f(t)$$

the integral of this function from time $t = t_0$ to time $t = t_1$ gives the total amount of insulin delivered over the period t_0 to $t1_1$.

Note that a flow rate can be measured in $\frac{\text{m}^3}{\text{s}}$. The integral of this value with respect to time would have units of m^3, or total volume.

8.5.4   Riemann Sums

One way to numerically approximate an integral over the range $a \le x \le b$ is to use Riemann Sums. The area under the curve is approximated by rectangles of a fixed width. The height of the individual rectangles are found by evaluating the function at numerous locations on the range $[a, b]$.

Given a function $f(x)$ to integrate between $x = a$ and $x = b$, N rectangles can be used. This means that the width of each rectangle h will be:

$$h = \frac{b - a}{N}$$

The area of each rectangle is given by:

$$f(x_i)\,h$$

where the value of x_i for the ith right Riemann sum is:

$$x_i = a + hi$$

and alternatively the value for x_i for the left Riemann sum is:

$$x_i = a + h(i - 1)$$

The area under the curve is approximated then as a sum of rectangle areas:

$$\int_a^b f(x)\,dx \approx \sum_{i=1}^{N} h\,f(x_i)$$

Note that integrals can be negative when the function is negative. Also, realize that the numerical approximation is a closer estimate of the true value for larger N, resulting in smaller values of h and numerous rectangles to evaluate.

8.5.5   Function Integration vs. Euler Integration

This chapter considers finding the area under a curve for a scalar function. The function value is known or can be determined at various positions along variable x. The methods presented can be readily extended for numerical integration of values across a surface or volume. Finding the area under the curve is obviously quite useful in many engineering applications where the total or average value is required.

For Ordinary Differential Equations, one numerical solution technique for initial value problems is called Euler Integration. For ODEs, the unknown value is a function of the dependent variable, like position and velocity values as functions of time in a particle dynamics problem. The value of the variable is not initially known for all points in time. Euler Integration allows for the next value to be determined from the current values. This will be discussed more in Chap. 14.

8.6   Trapezoidal Integration

Rather than use a rectangle to approximate the integral, a trapezoid can be used to approximate the area under the function. This requires evaluating the function at two points for each trapezoid. The area of an individual trapezoid is given by:

$$\frac{h}{2}\,(f(x_i) + f(x_{i+1}))$$

This means that the function is evaluated at the left and right side of the trapezoid. The value of x is given by:

$$x_i = a + h\,i$$

The overall integral is basically:

$$\int_a^b f(x)\, dx \approx \sum_{i=1}^{N} \frac{h}{2} \left(f(x_{i-1}) + f(x_i) \right)$$

This can be simplified, realizing that the right side of trapezoid i is the left side of trapezoid $i + 1$. As a result, the simplified version of the trapezoidal rule for integration is:

$$\int_a^b f(x)\, dx \approx \frac{h}{2} \left(f(x_0) + 2 \sum_{i=1}^{N-1} f(x_i) + f(x_N) \right)$$

Note that the value for x_0 is a and the value for x_N is b. The middle terms are doubled because the "inner" sides of the trapezoids will appear twice.

Example 8.5 (Average Flow) You are told that the flow rate from a reactor is given by the expression:

$$F(t)\, (\text{gpm}) = 10e^{-t/6}$$

You are asked to find the average flow from $t = 0$ to $t = 12$ using the trapezoidal rule with $N = 3$. Since $N = 3$, $a = 0$, and $b = 12$, h can be found as:

$$h = \frac{b - a}{N} = \frac{12 - 0}{3} = 4$$

This means the function must be evaluated at time points $t = 0, 4, 8,$ and 12. The following table shows the function values:

i	t (min)	$F(t)$ (gpm)
0	0	10.00
1	4	5.13
2	8	2.64
3	12	1.35

With $N = 3$ there are three trapezoids. The formula can be evaluated as:

$$\int_a^b f(x)\, dx \approx \sum_{i=1}^{N} \frac{h}{2} \left(f(x_{i-1}) + f(x_i) \right)$$

$$= \frac{4}{2} (10.00 + 5.13) + \frac{4}{2} (5.13 + 2.64) + \frac{4}{2} (2.64 + 1.35)$$

$$= 53.8\, (\text{gallons})$$

Using the simplified formula:

$$\int_a^b f(x)\, dx \approx \frac{h}{2}\left(f(x_0) + 2\sum_{i=1}^{N-1} f(x_i) + f(x_N) \right)$$

$$= \frac{4}{2}\,(10.00 + 2\,(5.13 + 2.64) + 1.35)$$

$$= 53.8 \text{ (gallons)}$$

The exact value is given as:

$$\int_0^{12} 10 e^{-t/6}\, dt = \left. -60 e^{-t/6} \right|_{t=0}^{t=12}$$

$$= (-8.12) - (-60) = 51.88$$

The units for the integral are gallons, since $F(t)$ has units of gallons/minute and t has units of minutes. To find the average flow rate, divide the integral by the total amount of time:

$$F_{avg} = \frac{53.8 \text{ gallons}}{12 \text{ min}} = 4.48 \text{ (gpm)}$$

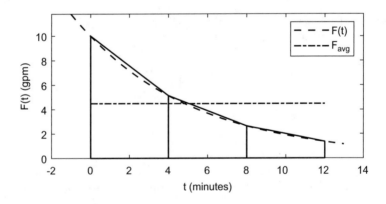

Using the trapezoidal rule for integration results in more accurate approximations that the use of Riemann sums. Similar to Riemann sums, using more trapezoids also results in more accurate approximations.

Problems

8.1 Given the equation:

$$y = 15e^{-2x} - x$$

Use Newton's method to find the value of x such that $f(x) = 0$ assuming $x_1 = 3$ as the initial guess. What is the function $f(x)$? What is $f'(x)$? Make a table showing x, $f(x)$, and $f'(x)$ for each iteration.

8.2 Given the equation:

$$y = 2x^3 + x^2 - 2x$$

Use Newton's method to find the value of x such that $f(x) = 0$ assuming $x_1 = 2$ as the initial guess. What is the function $f(x)$? What is $f'(x)$? Make a table showing x, $f(x)$, and $f'(x)$ for each iteration. Repeat this exercise starting with $x_1 = -2$. Do you get the same solution? Why?

8.3 Given the equation:

$$y = 2x - e^{-3x}$$

Use Newton's method to find the value of x such that $f(x) = 0$ assuming $x_1 = 4$ as the initial guess. What is the function $f(x)$? What is $f'(x)$? Make a table showing x, $f(x)$, and $f'(x)$ for each iteration.

8.4 Given the equation:

$$y = \sin(x) + 2x + 3$$

Use Newton's method to find the value of x such that $f(x) = 0$ assuming $x_1 = 5$ as the initial guess. What is the function $f(x)$? What is $f'(x)$? Make a table showing x, $f(x)$, and $f'(x)$ for each iteration.

8.5 Given the equation:

$$y = \cos(x) + x^2 - 2x$$

Use Newton's method to find the value of x such that $f(x) = 0$ assuming $x_1 = 0$ as the initial guess. What is the function $f(x)$? What is $f'(x)$? Make a table showing x, $f(x)$, and $f'(x)$ for each iteration. Repeat this exercise starting with $x_1 = 3$. Do you get the same solution? Why?

8.6 Given the equation:

$$y = f(x) = x^2$$

Use Newton's method to find the value of x such that $f(x) = 2$ assuming $x_1 = 3$ as the initial guess. Show your values for x, $f(x)$, and $f'(x)$ at each iteration.

8.7 The discharge of a pacemaker battery follows a known function of time. The battery voltage must not drop below a limiting value of 2 volts. The voltage follows the equation:

$$V(t) = 3e^{-2t} + (1 - t^2)$$

Use Newton's method to find the time when $V(t) = 2$ assuming $t_1 = 1$ as the initial guess. What is the function $f(t)$? What is $f'(t)$? Make a table showing t, $f(t)$, and $f'(t)$ for each iteration. You may want to plot $V(t)$ as a function of time to determine approximately when $V(t) = 2$.

8.8 Given the equation:

$$y = f(x) = x^4 + 2e^{3x}$$

Find the value of x such that $f(x) = 4$ using Newton's method: assume $x_1 = 1$ as the initial guess and find x_4. Make a table showing x, $f(x)$, and $f'(x)$ for each iteration.

8.9 Given the equation:

$$g(x) = x^2 \quad h(x) = 2x^3$$

Find the value of x such that $g(x) = h(x)$ using Newton's method. What is the $f(x)$ used to solve in the form $f(x) = 0$? Assume $x_1 = 2$ as the initial guess and find x_4. Make a table showing x, $f(x)$, and $f'(x)$ for each iteration.

8.10 Given the equation:

$$g(x) = \frac{1}{(2x + 1)^3} \quad h(x) = 2x^3$$

Find the value of x such that $g(x) = h(x)$ using Newton's method. What is the $f(x)$ used to solve in the form $f(x) = 0$? Assume $x_1 = 2$ as the initial guess and find x_4. Make a table showing x, $f(x)$, and $f'(x)$ for each iteration.

8.11 Given the equation:

$$y = x^2 - 2$$

Use trapezoidal integration to integrate the function from $x = 0$ to $x = 3$ using $N = 9$. You may want to make a table to show your values.

8.12 Given the equation:

$$y = x^3 + e^x$$

Use trapezoidal integration to integrate the function from $x = 0$ to $x = 4$ using $N = 6$. You may want to make a table to show your values.

8.13 Given the equation:

$$y = \frac{100}{(2x + 1)^3}$$

Use trapezoidal integration to integrate the function from $x = 1$ to $x = 3$ using $N = 8$. You may want to make a table to show your values.

8.14 Given the equation:

$$y = x^4$$

Use trapezoidal integration to integrate the function from $x = -1$ to $x = 3$ using $N = 8$. You may want to make a table to show your values.

8.15 Given the equation:

$$y = 2 + \cos x + x$$

Use trapezoidal integration to integrate the function from $x = 0$ to $x = \pi$ using $N = 8$. You may want to make a table to show your values.

Further Reading

1. Chapra, S., & Canale, R. (2020). *Numerical methods for engineers* (8th ed.). McGraw-Hill Education.
2. Davis, R.A. (2021). *Practical numerical methods for chemical engineers: Using Excel with VBA* (5th ed.).
3. Law, V.J. (2013). *Numerical methods for chemical engineers using Excel, VBA, and MATLAB* (1st ed.). CRC Press.
4. Moin, P. (2010). *Fundamentals of engineering numerical analysis* (2nd ed.). Cambridge University Press.
5. Rao, S.S. (2001). *Applied numerical methods for engineers and scientists* (1st ed.). Pearson.

Chapter 9
Introduction to Numerical Optimization

9.1 The Objective Function

Numerical optimization methods typically assume that one can calculate a scalar value that is to be maximized or minimized. This is considered the *cost function* or the *objective function*. Generally, the cost function is a mathematical function of *decision variables* or *unknowns*. In many introductory calculus classes, a function of a single variable is minimized or maximized by finding the critical points where the first derivative is equal to zero. In more advanced calculus classes, a function of two variables can be minimized by finding points that make the gradient equal to zero. These concepts extend directly to more complicated cases.

In many engineering problems, you could have numerous decision variables. In order to simplify things, these unknowns can be stacked together using vector notation. You are not limited to two or three unknowns in a vector value. Typically, our vector with n unknown values will be written simple as $\underline{x}$, with $\underline{x} \subset \mathbb{R}^n$. This means that $\underline{x}$ is some point in an n-dimensional space. Usually, we will know bounds on $\underline{x}$, so this means our variables $\underline{x}$ belong to the set of points X, which can also be written as $x \in X \subset \mathbb{R}^n$.

For example, in a chemical plant design problem, the size of four reactors could dictate the overall plant cost:

$$\underline{x} = \begin{bmatrix} V_1 \\ V_2 \\ V_3 \\ V_4 \end{bmatrix}$$

Let us define the cost function as a function of the decision variables $\underline{x}$ as $f(\underline{x})$. We will assume that we only have a single cost function, so this function f can be thought of as a mapping from any point in the set X to a number, $f : X \rightarrow \mathbb{R}$. An optimization routine must search the allowable solution space of the decision

E. Gatzke, *Introduction to Modeling and Numerical Methods for Biomedical and Chemical Engineers*, https://doi.org/10.1007/978-3-030-76449-4_9

variables to find the best value of the objective function that satisfies the problem constraints. The general mathematical form of the problem could be written as:

$$\min f(\underline{x})$$

subject to constraints on $\underline{x}$

There are many different objective functions one could seek to minimize or maximize. Some examples include the following:

- The distance from a point in 2D or 3D space with variables x, y, and z. $\underline{x} = [x\ y\ z]^T$.
- The sum square error for a model as compared to actual measurements with model parameters as variables.
- The total distance traveled between two points, with variables indicating whether or not a path was taken.
- The total cost for a chemical plant with the number and size of reactors as variables.
- The value of a stock portfolio with different investment options as variables.
- The overall strength of a composite material with individual material quantities as variables.

Variables can be continuous like the size and temperature of a reactor or the position in space. In other cases, variables can take binary values to indicate whether or not a certain action is performed, 0=no and 1=yes. In other cases, variables take only integer values, like the number of reactors in a chemical plant. Binary and integer variables complicate things in many cases. This will be considered in more detail later.

Additionally, some problems require minimization, while others require maximization. Minimizing $f(\underline{x})$ is the same as maximizing $-f(\underline{x})$. We will only discuss minimization problems in the current chapter.

9.2 Unconstrained Optimization

The general form for an unconstrained optimization problem can be written as:

$$\min f(\underline{x})$$

The critical points for this function occur when the gradient of $f(\underline{x})$ is equal to 0, $\frac{\partial f}{\partial x}(\underline{x}) = 0$. The gradient of the objective function, $\frac{\partial f}{\partial x}(\underline{x}) = \nabla f(\underline{x})$, is an n-dimensional vector function of $\underline{x}$. For example, say your cost function maps points in a two-dimensional space to a cost:

$$f(\underline{x}) = x^3 y^2 + x^2 y^4 + 5x + 7y$$

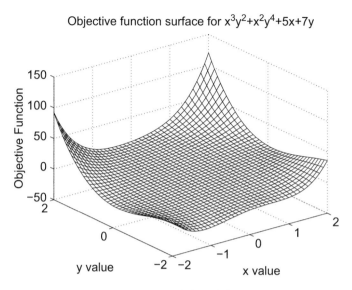

Fig. 9.1 Objective function surface for $f(\underline{x}) = x^3 y^2 + x^2 y^4 + 5x + 7y$

The gradient of this function is:

$$\nabla f(\underline{x}) = \begin{bmatrix} \frac{\partial f}{\partial x} \\ \frac{\partial f}{\partial y} \end{bmatrix} = \begin{bmatrix} 3x^2 y^2 + 2xy^4 + 5 \\ 2x^3 y + 4x^2 y^3 + 7 \end{bmatrix}$$

To find a point that minimizes this function, one would have to solve a fairly complex set of nonlinear algebraic equations. Obviously, this becomes difficult for even simple functions involving only two variables. You would have n different functions to solve in order to make the n different partial derivatives equal to zero for a problem with n variables (Fig. 9.1).

The gradient vector for $f(\underline{x})$ is obviously very important for the determination of the final solution. Given any point $\underline{x}_0$ in X, the gradient vector points in the direction of steepest increasing value of $f(\underline{x})$ from the point $\underline{x}_0$. One could imagine a hill-climbing method for maximization problems or a steepest descent method for minimization. For minimization, from a starting point $\underline{x}_0$, one could perform iterative search looking for $\nabla f(\underline{x}) = 0$ using the formula $\underline{x}_{new} = \underline{x}_{old} - K \frac{\partial f}{\partial x}(\underline{x}_{old})$. For some values of K, this numerical method may be unstable. Remember that the result may only be a local optimum, and there may be better solutions in other parts of the solution space. Additionally, care should be taken to avoid saddle points. Second derivative information can provide additional information about the shape of the objective function at a point in the parameter space.

9.3   Constraints

In many problems, additional functions are used to limit the solution space. Constraints arise from the model equations of the problem. Variable bounds can be seen as constraints that limit the variable values to a limited region in the n-dimensional space, $\mathbb{R}^n$. For example, the variable for product inventory level during a given time period could be related to the previous inventory level, the number of deliveries received over the period, and the number of orders sent during that period, resulting in the constraint:

$$x_{t+1} = x_t + d_t - o_t$$

A general form for the constrained optimization problem is:

$$\min f(\underline{x})$$

$$\text{subject to } g(\underline{x}) \le 0$$

with the constraint functions $g : X \to \mathbb{R}^m$ for m separate constraints. This means that the function g maps the points in the set X to an m-dimensional space. Inequality constraints are like fences that keep the variable values inside a specified region.

Consider the case where $g(\underline{x})$ is a single constraint, a scalar function of the vector $\underline{x}$. The problem can be written using a Lagrangian relaxation to form an unconstrained problem. The new problem becomes:

$$\min f(x) + \lambda g(x)$$

where λ is a non-negative value, $\lambda \ge 0$. This can be seen as a penalization for violating the constraint $g(\underline{x}) \le 0$. When $g(\underline{x})$ is positive, the objective function increases, so it is desirable to have $g(\underline{x})$ be negative. The unconstrained optimization problem can be solved iteratively, changing the values of λ until the minimum value for λ is found that still keep $g(\underline{x}) \le 0$.

9.3.1   Smallville Example

You have decided to take a job in Smallville. You are searching for a house inside the city limits, but you would like to find a house close to your job in order to minimize your commute distance and time. You can assume that house locations in Smallville are denoted by an $x \times y$ position with $0 \le x \le 3$ and $0 \le y \le 2$. Your business location is at $x = 4$, $y = 3$. There is a dump at the $x = 0$, $y = 1$ location, and you can smell it for at least 1.5 miles. These constraints are shown graphically in Fig. 9.2.

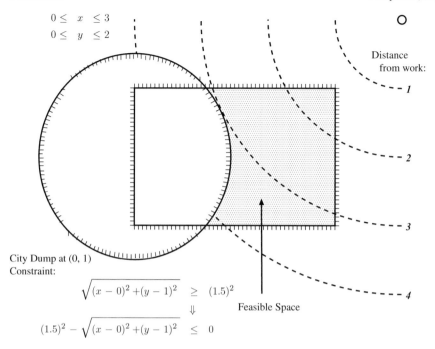

Town limits:

$0 \le x \le 3$
$0 \le y \le 2$

Workplace: (4, 3)

Distance from work:

1

2

3

City Dump at (0, 1)
Constraint:

$$\sqrt{(x-0)^2 + (y-1)^2} \ge (1.5)^2$$
$$\Downarrow$$
$$(1.5)^2 - \sqrt{(x-0)^2 + (y-1)^2} \le 0$$

Feasible Space

4

Fig. 9.2 Constraints and objective function contours for Smallville example

The solution to this problem is pretty obvious, but we should try to formulate the problem in the standard form. Our variables are x and y, so our vector of unknowns is $\underline{x} = [x \ y]^T$. The distance from work is $\sqrt{(x-4)^2 + (y-3)^2}$. There are limits on x and y based on the city limits. The dump poses an interesting constraint, such that x and y must be at least 1.5 units away from the (0, 1) position. This constraint can be written in the form: $\sqrt{(x-0)^2 + (y-1)^2} \ge 1.5$. The problem becomes:

$$\min \quad \sqrt{(x-4)^2 + (y-3)^2}$$
$$s.t. \quad 0 \le x \le 3$$
$$0 \le y \le 2$$
$$\sqrt{(x-0)^2 + (y-1)^2} \ge 1.5$$

We can now identify $f(\underline{x})$ and $g(\underline{x})$ for this problem. The distance from work is:

$$f(\underline{x}) = \sqrt{(x-4)^2 + (y-3)^2}$$

To simplify this objective function a bit, you can try to minimize the distance from work squared instead of minimizing the distance.

$$f(\underline{x}) = (x - 4)^2 + (y - 3)^2$$

There are five constraints for this problem that help limit our solution space:

$$g(\underline{x}) = \begin{cases} -x & \leq 0 \\ x - 3 & \leq 0 \\ -y & \leq 0 \\ y - 2 & \leq 0 \\ 1.5 - \sqrt{(x - 0)^2 + (y - 1)^2} & \leq 0 \end{cases}$$

The overall problem can now be formulated in a standard form with five constraints:

$$\begin{aligned} \min \quad & (x - 4)^2 + (y - 3)^2 \\ & -x \leq 0 \\ & x - 3 \leq 0 \\ & -y \leq 0 \\ & y - 2 \leq 0 \\ & 1.5 - \sqrt{(x - 0)^2 + (y - 1)^2} \leq 0 \end{aligned}$$

The solution in this case is pretty clear: the point $x = 3$, $y = 2$ is the closest point to work that is in the city limits and outside the range of the dump. This point satisfies exactly constraints 2 and 4. To be satisfied exactly means that $g_i(\hat{x}) = 0$. The other constraints have a value less than 0, $g_i(\hat{x}) < 0$.

9.4 Optimality Conditions

Once you have a feasible point that does not violate any of your constraints, you can check that point to see if it may be a locally optimum solution. Usually, for constrained optimization problems, local solutions occur where some of the inequality constraints are exactly satisfied. For constraint $g_i(x) \leq 0$, the constraint is exactly satisfied at the point $\hat{x}$ if $g_i(\hat{x}) = 0$. These constraints are called "active" or "binding" constraints. Equality constraints must always be satisfied, so they should always be active. Once you have a point in your feasible space, you can check for optimality using conditions developed by Karush, Kuhn, and Tucker, the KKT conditions. Note that these conditions do not say anything about how to find a KKT point; instead, they just give you a test for a given point.

For many years, these conditions were just KT conditions, as they were originally published by Kuhn and Tucker. Eventually, it was discovered that an obscure Indian mathematician, Karush, had already discovered them so now we just call them KKT conditions.

KKT Conditions

For a potential solution $\hat{x}$, the following conditions hold. The set I specifies the binding constraints at the point $\hat{x}$, $I = \{i : g_i(\hat{x}) = 0\}$. Binding constraints are satisfied exactly at $\hat{x}$. Additionally, $\nabla g_i(\hat{x})$ should be linearly independent. If the following conditions hold at $\hat{x}$, then $\hat{x}$ is a KKT point and a local solution.

$$-\nabla f(\hat{x}) = \sum_{i \in I} \lambda_i \nabla g_i(\hat{x})$$

$$\lambda_i \geq 0$$

Note that this does not specify how to find a KKT point.

9.4.1 Optimality of Smallville Example

From our Smallville problem, we think the point (3, 2) is the solution. You know that constraints 2 and 4 are satisfied at this point. You can find the gradient of these constraints and the gradient of the objective function at this point, (3, 2). The KKT conditions basically say that you must be able to find positive multipliers for the gradient directions of the active constraints that will add the active constraint directions together to get the gradient of the objective function improving direction. These active constraints at (3, 2) are:

$$x - 3 \leq 0$$

$$y - 3 \leq 0$$

The gradients of these two constraints are:

$$\nabla g_2 = \begin{bmatrix} 1 \\ 0 \end{bmatrix} \quad \nabla g_4 = \begin{bmatrix} 0 \\ 1 \end{bmatrix}$$

With the negative gradient of the objective function at (3, 2) being $[2\,2]^T$, the KKT conditions at (3,2) are:

$$\begin{bmatrix} 2 \\ 2 \end{bmatrix} = \lambda_1 \begin{bmatrix} 1 \\ 0 \end{bmatrix} + \lambda_2 \begin{bmatrix} 0 \\ 1 \end{bmatrix}$$

$$\lambda_1 \geq 0$$

$$\lambda_1 \geq 0$$

In the case of the point $(3, 2)$, $\lambda_i = 2$ for both constraints 2 and 4. The λ values are positive, so the point $(3, 2)$ is a KKT point or a local solution.

Now, consider optimality conditions at the point $(3, 0)$. This point also has two active constraints, constraints 2 and 3. The KKT conditions at the point $(3, 0)$ with two active constraints are:

$$\begin{bmatrix} 2 \\ 6 \end{bmatrix} = \lambda_1 \begin{bmatrix} 1 \\ 0 \end{bmatrix} + \lambda_2 \begin{bmatrix} 0 \\ -1 \end{bmatrix}$$

$$\lambda_1 \geq 0$$

$$\lambda_1 \geq 0$$

There are no positive values for λ that will resolve the gradients of the active constraints in the direction of $-\nabla f$. The gradient of the two active constraints dictates two directions, and assuming λ_i must all be positive multipliers for these two directions, you end up with a cone of points expanding out to the right and down in the figure. The improving direction $-\nabla f$ must lie in this cone of directions, but at the point $(3, 0)$, the improving direction does not lie in that cone. Therefore, the point $(3, 0)$ will not be a KKT point. Solving the set of two linear algebraic equations from the KKT conditions, $\lambda_1 = 2$ for constraint 2 and $\lambda_2 = -6$ for constraint 3 (Fig. 9.3).

9.4.2 Equality Constraints and Infeasibility

Equality constraints may restrict the feasible region. Equality constraints can be written as two inequality constraints.

$$g_i(\underline{x}) = 0$$

$$\Downarrow$$

$$g_i(\underline{x}) \leq 0$$

$$g_i(\underline{x}) \geq 0$$

In the Smallville example, you may want to locate your house on Main St. In this problem, the locations on Main St. correspond to:

$$y = x^2 - 1.5$$

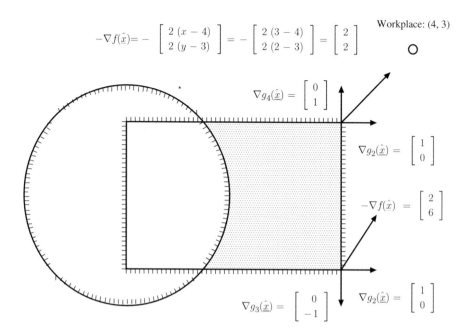

$$-\nabla f(\hat{x}) = -\begin{bmatrix} 2(x-4) \\ 2(y-3) \end{bmatrix} = -\begin{bmatrix} 2(3-4) \\ 2(2-3) \end{bmatrix} = \begin{bmatrix} 2 \\ 2 \end{bmatrix}$$

Workplace: (4, 3)

$$\nabla g_4(\hat{x}) = \begin{bmatrix} 0 \\ 1 \end{bmatrix}$$

$$\nabla g_2(\hat{x}) = \begin{bmatrix} 1 \\ 0 \end{bmatrix}$$

$$-\nabla f(\hat{x}) = \begin{bmatrix} 2 \\ 6 \end{bmatrix}$$

$$\nabla g_3(\hat{x}) = \begin{bmatrix} 0 \\ -1 \end{bmatrix}$$

$$\nabla g_2(\hat{x}) = \begin{bmatrix} 1 \\ 0 \end{bmatrix}$$

Fig. 9.3 Active constraint gradients and objective function improving direction for Smallville example at solution (3, 2) and the point (3, 0)

This can introduce two new constraints to the problem:

$$g_6(x) = y - x^2 + 1.5 \le 0$$
$$g_7(x) = -y + x^2 - 1.5 \le 0$$

This means that the feasible set of points are only the ones along the equality constraint. Instead of a fence boundary as in inequality constraints, equality constraints dictate a single "street" that points can move along when searching for optimality. The resulting solution would be at the point ($\sqrt{3.5}$, 2). This point is inside the rectangular box of constraints dictating the size of the town, more than 1.5 miles from the dump location, and exactly on Main St. Constraints 4, 6, and 7 would be active. At the point ($\sqrt{3.5}$, 2), the following KKT conditions could be considered:

$$\begin{bmatrix} -2(\sqrt{3.5}-4) \\ 2 \end{bmatrix} = \lambda_1 \begin{bmatrix} 0 \\ 1 \end{bmatrix} + \lambda_2 \begin{bmatrix} -2\sqrt{3.5} \\ 1 \end{bmatrix} + \lambda_3 \begin{bmatrix} 2\sqrt{3.5} \\ -1 \end{bmatrix}$$

$$\lambda_1 \ge 0$$
$$\lambda_2 \ge 0$$
$$\lambda_3 \ge 0$$

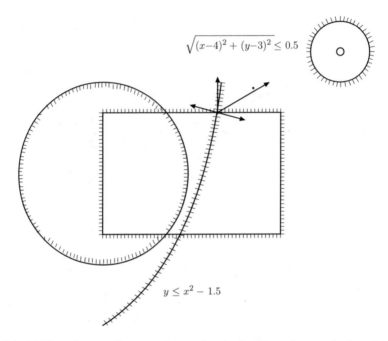

Fig. 9.4 Addition of an equality constraint restricts the feasible region to only the points that exactly satisfy the equality constraint. In this case, $y = x^2 - 1.5$. Addition of the constraint $\sqrt{(x-4)^2 + (y-3)^2} \leq 0.5$ makes the problem infeasible, as no points can satisfy all the constraints

Constraints 4 and 6 define a cone of directions that contains the direction $-\nabla f$, so positive multipliers for λ_1 and λ_2 can be found with $\lambda_3 = 0$ to satisfy the KKT conditions.

Perhaps we would like to limit our search to locations within half a mile. The resulting inequality constraint is:

$$\sqrt{(x - 4)^2 + (y - 3)^2} \leq 0.5$$

Obviously, we now have too many constraints and no feasible points exist. The problem can be infeasible if no points can be found that satisfy the problem constraints (Fig. 9.4).

9.5 Convexity

At this point, issues involving convexity of sets and multivariable functions should be addressed. A convex set X satisfies the relationship $\lambda x_1 + (1 - \lambda)x_2 \in X$ for all $0 \leq \lambda \leq 1$, $\forall x_1, x_2 \in X$. This just means that given a set of points X, you can draw a line between any two points in the set and all points on the line will still be in the set. This definition is not especially useful, since we usually deal with functions instead of sets.

A convex set can be constructed from a convex function by evaluating the epigraph of a convex function. If $x \in X \subset \mathbb{R}^n$, $f : X \rightarrow \mathbb{R}$, $epi(f) \in \mathbb{R}^{n+1}$. These are all the points "above" the function. In our example, you can imagine the objective function mapping values of x and y to a surface in 3D space. The points above this surface make up a set of points. If the function is convex, the set of points in the epigraph will be convex.

As in convexity results from calculus, convexity of a function requires analysis of a second-order condition. The Hessian matrix H can be calculated for $f(x)$ as:

$$H(x) = \begin{bmatrix} \frac{\partial^2 f}{\partial x_1 \partial x_1} & \frac{\partial^2 f}{\partial x_1 \partial x_2} & \cdots & \frac{\partial^2 f}{\partial x_1 \partial x_n} \\ \frac{\partial^2 f}{\partial x_2 \partial x_1} & \frac{\partial^2 f}{\partial x_2 \partial x_2} & & \vdots \\ \vdots & & \ddots & \vdots \\ \frac{\partial^2 f}{\partial x_n \partial x_1} & \cdots & \cdots & \frac{\partial^2 f}{\partial x_n \partial x_n} \end{bmatrix}$$

A given function $f(x)$ is convex if the Hessian of the function is positive semi-definite. This means that all the eigenvalues of $H(x)$ are ≥ 0. Note that the Hessian matrix may be a function of x, making the calculation of the eigenvalues more difficult. You could evaluate the Hessian at a single point to determine the convexity of the function at that single point. For optimization problems, x belongs to a set of possible feasible points. Some methods do exist to bound the smallest eigenvalue for a general Hessian matrix using interval analysis methods. In general, determining if a general nonlinear function is convex over a region x can be quite difficult.

Why is convexity of a function important? If all the constraint functions and the objective function are convex and there is a feasible point in the solution space, there is a single solution to the problem. This single solution could be degenerate, meaning multiple points in X result in the same objective function value. If a problem involves nonconvex constraint or objective functions, simple solution methods can only guarantee local solutions.

Given the function $f(x) = (x_1)^3 + x_2$, the gradient is determined by $\nabla f = \begin{bmatrix} 2x_1^2 \\ 1 \end{bmatrix}$, and the 2×2 Hessian matrix is given by $H(x) = \begin{bmatrix} 4x_1 & 0 \\ 0 & 0 \end{bmatrix}$. Since the Hessian in this case is a diagonal matrix, the eigenvalues are known to be $4x_1$ and 0. Note that the minimum eigenvalue value depends on the range of values for x_1. If the lower bound on x_1 is positive, the minimum eigenvalue is 0. This makes the Hessian positive semi-definite and the function convex over the range of x. If the lower bound on x_1 is < 0, the Hessian is not semi-definite and the function is nonconvex over the range of x.

Linear equality and inequality constraints are convex. Nonlinear inequality constraints may be convex. Nonlinear equality constraints are always nonconvex. A nonlinear function $f(x) = 0$ can be written as two inequality constraints:

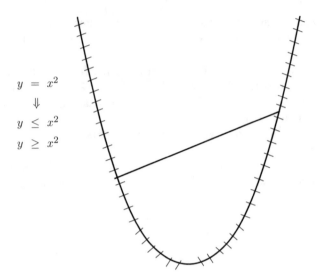

$$y = x^2$$
$$\Downarrow$$
$$y \le x^2$$
$$y \ge x^2$$

Fig. 9.5 The nonconvex set $y = x^2$

$$0 \le f(\underline{x}) \le 0$$

This implies that if $f(\underline{x})$ is nonlinear and convex at some point then one of the two following inequality constraints would be nonconvex at the same point:

$$0 \le f(\underline{x})$$
$$f(\underline{x}) \le 0$$

since this is the same as:

$$-f(\underline{x}) \le 0$$
$$f(\underline{x}) \le 0$$

Therefore one constraint must be nonconvex at the point in question. Think of the equation of a line, $y = x^2$; this is shown in Fig. 9.5. The set of points that satisfy this equality constraint are only the points on the line. Now, consider only the points defined by $y \ge x^2$, the points "above" the line. This single inequality would define a convex set of points. Putting the inequality constraint in our general form $x^2 - y \le 0$, our constraint becomes $g(\underline{x}) = x^2 - y \le 0$. This constraint generates a convex set of points in the $x \times y$ space. The function x^2 is convex for all values x, $x \in \mathbb{R}^1$, since the Hessian is the 1×1 matrix [2]. The epigraph of x^2 defines the convex set of points for y such that $y \ge x^2$, the convex set of points in $\mathbb{R}^2$, $x \times y$.

Consider the constraint for all points in a sphere of radius r, where r is some constant value:

$$x^2 + y^2 + z^2 \leq r^2$$

The Hessian of the constraint function is:

$$\begin{bmatrix} 2 & 0 & 0 \\ 0 & 2 & 0 \\ 0 & 0 & 2 \end{bmatrix}$$

This implies that the eigenvalues are all positive with a value of 2. We know intuitively that the points in a sphere should be a convex set. Flipping the sign on the inequality would examine the points outside the sphere of radius r. The "Swiss cheese" set is nonconvex, as it is easy to imagine a line between two points containing points inside the empty sphere.

9.6 Types of Optimization Problems

Linear Programming (LP) In some cases, the general optimization form has linear objective functions and linear constraints. An optimization problem can be found in the form:

$$\min Cx$$

$$\text{subject to } Ax \leq b, lb \leq x \leq ub$$

This is a special linear constrained case where the objective function is a linear function of the decision variable vector x and the constraints are also linear. This can readily be solved using the *linprog* command in the MATLAB Optimization Toolbox, even for large-scale problems (hundreds or thousands of variables and constraints). Generally, the simplex method is used, but interior point methods are gaining popularity and speed of calculation.

Quadratic Programming (QP) For cases with a convex quadratic objective of the form:

$$\min \frac{1}{2} x^T H x + Cx$$

$$\text{subject to } Ax \leq b, lb \leq x \leq ub$$

The problem is termed a Quadratic Program (QP). The MATLAB command *quadprog* can be used to solve this type of problem.

Nonlinear Programming (NLP) General cases with nonlinear functions in the objective function and constraint functions take the form:

$$\min f(x)$$

$$\text{subject to } g(x) = 0, \ h(x) < 0, lb \leq x \leq ub$$

The problem is termed a Nonlinear Program (NLP). The MATLAB command *fmincon* can be used to solve this type of problem. If nonlinear equality constraints are included, the problem is definitely nonconvex and local solution may be encountered. In special cases,

$$\min f(x)$$

$$\text{subject to } Ax \leq b, \ h(x) < 0, lb \leq x \leq ub$$

if the functions $f(x)$ and $h_i(x)$ are all convex over the range of x, the problem is a convex NLP and has a single solution.

Mixed Integer Linear Programming For problems where some variables can only take binary values, the problem is considered a mixed integer problem. A common mixed integer problem is the Mixed Integer Linear Programming (MILP) problem of the form

$$\min Cx$$

$$\text{subject to } Ax \leq b, lb \leq x \leq ub, x_i \in \{0, 1\}$$

There exist specialized methods for solving problems where the decision variables are only allowed to take values of 0 or 1 rather than values between 0 and 1 (inclusive). The MATLAB Optimization Toolbox includes the *intlinprog* function that can solve some MILP problems.

Problems

9.1 For the following objective function, determine the improving direction:

$$f(x, y) = 3x - 4y$$

Is the function linear or nonlinear?
Is the function convex or nonconvex? Why?

9.2 For the following objective function, determine the improving direction:

$$f(x, y) = 2x^2 - 4y^2$$

Is the function linear or nonlinear?
Is the function convex or nonconvex? Why?

9.3 For the following constrain function, put it in standard form:

$$x^2 = y^2 + 4$$

Is the constraint linear or nonlinear?
Is the constraint convex or nonconvex? Why?

9.4 For the following constrain function, put it in standard form:

$$y \geq x^4 + 8$$

Is the constraint linear or nonlinear?
Is the constraint convex or nonconvex? Why?

9.5 For the following constrain function, put it in standard form:

$$y \geq 6 - 3x$$

Is the constraint linear or nonlinear?
Is the constraint convex or nonconvex? Why?

9.6 For the following optimization formulation:

$$\min x + y + z$$
$$2x - 5y = 5$$
$$z \leq x + 2$$
$$0 \leq x + y + z$$

Is the formulation linear or nonlinear?
Is the formulation convex or nonconvex? Why?
Is the problem an LP, QP, Convex NLP, Nonconvex NLP, MILP, or something else?
Justify your answer.

9.7 For the following optimization formulation:

$$\min x^3 + y^3$$
$$-4 \leq x$$
$$-2 \leq y$$

Is the formulation linear or nonlinear?
Is the formulation convex or nonconvex? Why?
Is the problem an LP, QP, Convex NLP, Nonconvex NLP, MILP, or something else?
Justify your answer.

9.8 For the following optimization formulation:

$$\min x + y$$
$$y = x^2$$
$$-5 \le y$$

Is the formulation linear or nonlinear?
Is the formulation convex or nonconvex? Why?
Is the problem an LP, QP, Convex NLP, Nonconvex NLP, MILP, or something else?
Justify your answer.

9.9 For the following optimization formulation:

$$\min x + y + z$$
$$y = 2x$$
$$x + y \le 2$$
$$x, y, z \in \{0, 1\}$$

Is the formulation linear or nonlinear?
Is the formulation convex or nonconvex? Why?
Is the problem an LP, QP, Convex NLP, Nonconvex NLP, MILP, or something else?
Justify your answer.

9.10 For the following optimization formulation:

$$\min x^2 + y^2$$
$$y = 2x + 4$$
$$4 \le x + y$$

Is the formulation linear or nonlinear?
Is the formulation convex or nonconvex? Why?
Is the problem an LP, QP, Convex NLP, Nonconvex NLP, MILP, or something else?
Justify your answer.

Further Reading

1. Bonnans, J.-F., Gilbert, J.C., Lemarechal, C., & Sagastizabal, C.A. (2006). *Numerical optimization: Theoretical and practical aspects* (2nd ed.). Springer.
2. Boyd, S., & Vandenberghe, L. (2004). *Convex optimization* (1st ed.). Cambridge University Press.

3. Hamming, R.W. (1987). *Numerical methods for scientists and engineers* (2nd ed.). Dover.
4. Nocedal, J., & Wright, S.J. (2006). *Numerical optimization* (2nd ed.). Springer.
5. Sierksma, G. (1987). *Linear and integer optimization: Theory and practice.* Chapman and Hall/CRC.

Chapter 10
Introduction to Spreadsheets

10.1 Spreadsheet Advantages and Disadvantages

The basic interface is quite simple. A matrix of data is presented, with rows labeled by increasing integer numbers and columns labeled with letters. Each cell of the matrix therefore can be referenced with a letter and number. The top left cell is A1. Numeric values can be put into cells. Strings and text can also be entered into cells (Fig. 10.1).

Spreadsheets are truly powerful in their ability to compute user derived functions from data. This is accomplished in spreadsheets by use of cell referencing. Specifically, relative referencing allows rapid computation of complex values. For example, consider the task of making a column of cells counting from 1 to 100. A user could sit and type all 100 numbers in the 100 cells. This task is more easily accomplished by putting the number 1 in cell A1, the formula =A1+1 in cell A2, then copying that cell down the column. Cell A2 contains the formula =A1+1 which resolves to the value of 2. Cell A3 will contain the formula =A2+1 which resolves to 3 since A2 has a value of 2. This list of numbers is rapidly calculated because each number depends on the number in the cell one cell above it. Additionally, the cells can be rapidly changed by changing one number. Rather than cells counting from 1 to 100, one could change all the cells to count from 101 to 200 just by changing the text in the first cell A1 from 1 to 101. All the cells below would update and change their values (Fig. 10.2).

Cell referencing is powerful, as formulas can be readily developed. By copying the cells, the relative reference is retained by default. However, in some cases the values in a copied list of formulas should reference a fixed cell location. Consider calculating compound interest. You have $100 so you put 100 in cell A1. Assuming 4% growth, you could put =A1*1.04 in the cell below, cell A2. Then you could copy that cell down to determine how much money you would have in ten years. Now, each formula explicitly has the value 1.04 in the equation, so to change the interest rate you would have to change all the equations. It would be more efficient to write

© Springer Nature Switzerland AG 2022
E. Gatzke, *Introduction to Modeling and Numerical Methods for Biomedical and Chemical Engineers*, https://doi.org/10.1007/978-3-030-76449-4_10

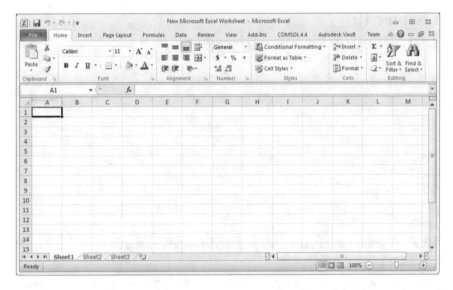

Fig. 10.1 Microsoft Excel interface. Data can be put into any of the cells, which are referenced by row and column. Formulas can also be put into cells to calculate new values based on the data

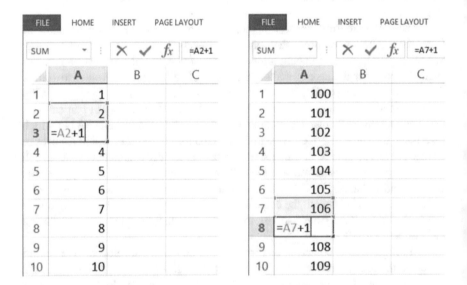

Fig. 10.2 Microsoft Excel interface showing a simple formula. Cell A1 contains a number. Each cell below A1 contains a formula that references the value of the cell above it. When cell A1 is changed, all the cells below it change as well

	A	B	C	D
1	100			
2	104			
3	=A2*1.04			
4	112.486			
5	116.986			
6	121.665			
7	126.532			
8	131.593			
9	136.857			
10	142.331			

SUM · X ✓ fx =A2*1.04

	A	B	C	D
1	100	1.04		
2	104			
3	108.16			
4	112.486			
5	116.986			
6	=A5*B1			
7	126.532			
8	131.593			
9	136.857			
10	142.331			

SUM · X ✓ fx =A5*B1

	A	B
1	500	1.05
2	525	
3	551.25	
4	578.813	
5	607.753	
6	638.141	
7	670.048	
8	703.55	
9	=A8*B1	
10	775.664	

SUM · X ✓ fx

Fig. 10.3 Microsoft Excel interface showing a simple formula with both relative referencing and fixed referencing. Cell A1 contains a number. Each cell below A1 contains a formula that references the value of the cell above. The second version shows that the formula also can include a fixed reference cell B1. When cells A1 or B1are changed, all the cells below it change as well

the equation in terms of a single cell reference. Putting the number 1.04 in cell B1, then writing the equation =A1*B1 in cell A2 would accomplish this task. Using the dollar sign fixes the cell reference. When you copy the contents of A2 into A3, the relative reference to the cell above remains relative, but the formula references the growth rate in cell B1. The formula in A3 would be =A2*B1 and the formula in A4 would be =A3*B1 resulting in the same numeric values in the column. However, now just by adjusting the value in cell B1, all the cells in the column will be updated (Fig. 10.3).

Spreadsheets are very good at organizing and calculating data. Spreadsheets include many mathematical functions. Spreadsheets also include mathematical functions that can operate over a range of cells such as SUM and AVERAGE. Spreadsheet applications also proved strong visualization tools for plotting and graphing data.

The biggest disadvantage of spreadsheet applications lies in the inherent limitations of the application for mathematical computations. While almost any calculation could be performed using a spreadsheet, some more complex calculations become quite difficult to perform. Rather than write a FOR loop or a WHILE loop to iterate during a calculation, complicated cell formulas involving fixed and relative references must be developed and copied to accomplish the same task. For very complex calculations, the cell referencing can become quite confusing, with formulas interacting with one another in very complex and hard to follow ways.

Engineers in industrial practice repeatedly have said that spreadsheet proficiency is one of the most important skills to develop while pursuing an undergraduate degree. Some engineers are quite capable at writing traditional calculations using IF, FOR, and WHILE constructs. Others are more capable at developing a spreadsheet calculation to accomplish the same task. Strong engineers should be able to accomplish the same task multiple different ways using whatever tools are available.

10.2 Excel Functions

Mastering the following commands will provide a basic level ability in use of Excel. Most of the commands below will also work in other spreadsheet applications, like Google Sheets.

- SUM—adds up a range of cells
- AVERAGE—takes the average of a range of cells
- MIN—determines the minimum value in a range of cells
- MAX—find the maximum value in a range of cells
- LN—natural log of a cell
- EXP—*e* to the power of the cell contents
- COUNT—counts the number of cells in a range that contain numbers
- IF—Checks a Boolean true/false condition and conditionally changes cell contents
- STDEV.S—computes the sample standard deviation value of a range of cells
- STDEV.P—computes the population sample standard deviation value of a range of cells

10.3 Tutorials

Links are provided to online tutorials so that the material can be rapidly updated when software changes are made in Excel.

10.3.1 Excel Tutorial 1

This tutorial presents basic introductory material on using Excel. Topics covered include inputting numeric data, inputting string data, creating a user-defined function, using cell referencing, and using relative cell referencing. The methods are presented as applied to use of Newton's method, trapezoidal integration, and Euler integration. http://cse.sc.edu/~gatzke/excel1.pdf

10.3.2 Excel Tutorial 2

This tutorial considers statistical functions in Excel. Simple statistics are computed, such as the mean and standard deviation. Values are plotted in a figure, and a trend line is added to the figure to show a function that fits the data along with its correlation coefficient. Error bars are added to a figure. From the Data Analysis Excel add-in, a linear regression is performed with additional information generated.

The Solver add-in is used to find the best-fit parameters to minimize the Sum Squared Error (SSE) of the measurement/model mismatch. Exponential decay data is transformed to a linear form, plotted, and evaluated. http://cse.sc.edu/~gatzke/excel2.pdf

10.3.3 Excel Tutorial 3

This tutorial examines the use of Visual Basic Application (VBA) code to generate an Excel macro function. With VBA, a user can write code with traditional flow control constructs including IF, FOR, and WHILE. When the VBA macro runs, it takes data from the Excel spreadsheet into variables inside the VBA macro. The macro can perform complex calculations, then write values into the Excel spreadsheet. Additionally, VBA can manipulate the Excel spreadsheet by deleting values and making figures. http://cse.sc.edu/~gatzke/excel3.pdf

10.3.4 Excel Tutorial 4

This tutorial shows how to filter data to remove unwanted data points. The sorting mechanism is used to organize data. A plot with error bars determined from the standard deviation of the data. The data is fit to an exponential model, then the data is transformed into a linear form. A trend line of the linear form data is used to determine model parameters. This tutorial is more open-ended than the first three, expecting the user to have retained concepts from the previous tutorials. http://cse.sc.edu/~gatzke/excel4.pdf

Problems

10.1 To make a column of time values starting at 0, then changing to 0.1, then 0.2 and so on. Assuming the numeric value 0 is in cell A1, what formula would go in cell A2 (including a relative reference) that could be copied down the column?

10.2 Assume that you want to evaluate the function $e^{-k_2 t}$. The value for k_2 is in cell B1 and the value for t is in cell B2. What formula would you put in cell B3 to evaluate this function using relative references to other cells?

10.3 Assume that time values are found in column A. Assuming that column B needs to contain a formula to evaluate the function $7e^{-t}$, what formula would go in cell B1 (including a relative reference) that could be copied down the column to quickly evaluate the function?

10.4 Assume that time values are found in column A. Assuming that column B needs to contain a formula to evaluate the function Ke^{-t}, where the parameter K is found in cell C1, what formula would go in cell B1 (including both a relative reference and a fixed reference) that could be copied down the column to quickly evaluate the function?

Further Reading

1. Davis, R.A. (2021). *Practical numerical methods for chemical engineers: Using Excel with VBA* (5th ed.).
2. Law, V.J. (2013). *Numerical Methods for Chemical Engineers Using Excel, VBA, and MATLAB* (1st ed.). CRC Press.

Chapter 11
Basic Probability and Statistics

11.1 Random Variables and Probability Distributions

Engineers like to work with equations that define the relationships between various constants and variables. Typically, we use mathematical methods to solve for the value of unknown variables. However, in many real cases, uncertainty may affect the equations. Values could take on a variety of different values. Randomness and variability can be understood by use of random variables.

> **Random Variables**—Random variables may take on any value from a set of possible outcomes.

The random variable does not have a single value. It may be characterized by the range of values, the type of values (typically integer or continuous), and the frequency of occurrence for values in the set. The randomness of the random variable is generally described by the probability distribution function for that variable.

> **Probability Distribution**—A probability distribution for a random variable defines the frequency of occurrence for all values in the possible set of outcomes.

The frequency of a random variable taking a value from the set depends on the probability distribution. The probability distribution function for a random variable is characterized by parameters to describe the probability of a random variable taking any specify variable from the set of all possible values for that variable.

Consider rolling an unbiased six-sided die. Each side has equal probability of occurring on any roll but one can never predict what the next outcome may be. However, understanding the randomness of the system can lead to reasoning about

© Springer Nature Switzerland AG 2022
E. Gatzke, *Introduction to Modeling and Numerical Methods for Biomedical and Chemical Engineers*, https://doi.org/10.1007/978-3-030-76449-4_11

the probable results. The set of possible outcomes are six different integer values: one to six. For any single roll, there is a one-in-six chance of rolling a value of six. Rolling two six-sided dice, there is a one in 36 chance of rolling two sixes simultaneously. The average value for any roll of a single die will approach 3.5 given enough samples.

> **Probability**—Probability is the prediction of future events related to the outcome of random variables.

Probabilistic reasoning is usually accomplished by study and analysis of various probability distributions. Probability answers the question "What could happen?" or "What are the chances that a specific event could happen?" The inverse problem is called *statistics* where data is used to make decisions.

> **Statistics**—Taking representative sampled measurements from a random system to determine useful information about the random system.

Probability allows us to predict that sum of four six-sided dice should *on average* be 14. Statistics would allow you to analyze repeated rolls of the dice to determine if they are truly unbiased.

Given a thorough understanding of the underlying physics leading to a random variable, one may develop a mathematical function describing the probability distribution for that random variable. In the example of the six-sided die, one assumes that each of the six sides have equal probability. No values below 1 or above 6 are possible. The resulting probability distribution function is called the discrete uniform distribution and may be written as:

$$f(x) = 1/6 \qquad \forall x \in \{1 : 6\}$$

The value x represents all the values in the possible set of outcomes. The function $f(x)$ defines the probability that any one value from the set occurs. Note that summing $f(x)$ over all possible set values (one to six) gives a value of 1. This represents 100% probability that one of the six values will occur any time this variable is sampled.

There are two initial ways to classify probability distributions:

- Discrete vs. continuous variables
- Variable range

Obviously, discrete variables take discrete (integer) values while continuous variables take any real vale. Similarly, the range of a variable is often limited by the type of problem considered. Some discrete random variable example include:

- The sum of three six-sided dice: range from 3 to 18
- The number of correct answers on a 50 question true/false test: range 0 to 50

- The number of machine breakdowns in one week: range 0 to N (repair time limits upper value of N)
- The quantity of broken items in a shipment of N items: range 0 to N
- The number of orders received on a single day: range 0 to ∞
- The number of employees hired or fired ($-$) in a given month: range $-N$ to $+\infty$ (actually limited to human population)

Note that given a specified range for a random variable, every value in that range has some nonzero probability of occurring. However, for many situations the probability is very low. One expects that is very rare to break every single item in a shipment or to receive millions of orders in a single day. These items could occur but the probability of such an occurrence approaches 0.

Some continuous random variables with ranges include:

- The concentration of a process stream: range 0 to C_{max} (Maximum depends on thermodynamics or measurement device reporting limit)
- The weight of a sample: range 0 to ∞ (Maximum actually limited to measurement device reporting limit)
- The change in temperature of an item after processing: range $-\Delta T_{min}$ to ΔT_{max}
- The pressure of a reactor: range 0 to ∞ (Maximum actually limited based on material strength)

In these cases, variable can take any of an infinite number of values that fall withing the range for the variable. However, the continuous variable when measured with modern equipment will often be reported with discretized values. Consider your body weight: your weight is a positive continuous value that ranges between 0 and some upper limit. However, your digital scale will generally report values with limited precision, perhaps one tenth of a pound. For any continuous value, we assume the actual real value is rounded to the closest discretized measurement device value.

11.2 Histograms

A histogram is often used to visualize the frequency distribution of a random variable.

> **Histogram**—A histogram graphically represents the frequency of occurrence for a random variable. From the set of variable values, multiple separate discrete bins are defined. The number of occurrences of the variable in each bin are plotted.

Given a sample of measurements for a random variable, a histogram represents the frequency of occurrence of different values of that variable. The histogram may also be derived from a known probability distribution for a variable. As the number

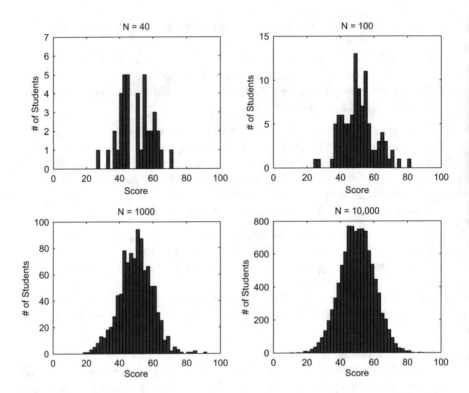

Fig. 11.1 Histograms generated from assuming student test scores are an average of 50 with a standard deviation of 10 randomly selected based on a normal distribution. Bins are 2 points wide. Note that even with 10,000 samples, the histogram does not necessarily form a perfectly smooth bell-shaped curve

of samples increases and the bin sizes become smaller, the histogram approaches the shape of the probability distribution function.

Consider the common use of representing student test score data using a histogram. A bin size is used to set the ranges of interest for the histogram. For a 100-point scale test, bins with a width of 10 points could be used. The number of student test scores in bin $90 < x \leq 100$ must be calculated, then the number of student test scores in the other bins is found. The data is presented in graphical form as a histogram. Note that the bin size and bin center placement can sometimes significantly influence the shape of the resulting histogram (Fig. 11.1).

11.3   Specific Probability Distributions

Probability distributions are mathematical representations of the probability of occurrence for a random variable. Many distributions can be fully characterized with only two parameters: the mean μ and the standard deviation σ. The mean represents the value a average of random variable should converge to given an infinite number of samples. The standard deviation represents the uncertainty or randomness for a variable.

This section will first discuss continuous selected probability distributions then introduce some discrete distributions. As stated previously, the probability distribution used to model a random variable should match the expected range of the variable. Additionally, the shape of the probability distribution should qualitatively match or be similar to the histogram for data plotted for a random variable.

11.3.1 Normal/Gaussian

The normal distribution probability density function is:

$$f(x,\ \mu,\ \sigma) = \frac{1}{\sigma\sqrt{2\pi}}\ e^{-\frac{(x-\mu)^2}{2\sigma^2}}$$

Plotting this function as a function of x for a known mean μ and standard deviation σ gives rise to a bell-shaped curve. Here, μ is the mean value and can take any value from $-\infty$ to ∞. The standard deviation σ must be non-negative. This value is related to the variance of a distribution which is defined as the squared standard deviation, σ^2. Note that the "tails" of the curve do extend out infinitely. This means that for this distribution, there is always a nonzero probability that any event could occur! The area under the curve always is exactly equal to 1. This means there is a 100% probability of something occurring.

The average value μ places the peak (center) of the normal distribution. The standard deviation σ dictates the "spread" of the data. From the center value, approximately 68% of the curve falls withing $\mu\pm\sigma$. This means that when sampling data from a normal distribution, approximately 68% of the samples should be withing the range $\mu - \sigma$ to $\mu + \sigma$. The exact value is found by integrating the density function from $x = \mu - \sigma$ to $x = \mu + \sigma$ resulting in a value of approximately 0.68269. Similarly, integrating from $x = -\infty$ to $x = \infty$ should result in the value 1.0, as stated previously (Fig. 11.2).

11.3.2 Continuous Uniform

Uniform means all events have the same probability of occurring. There is no "bell-shaped" curve for the probability distribution, is is just a constant between two integer values a and b. Any value between a and b has the same possible probability of occurrence. The probability distribution function is simply:

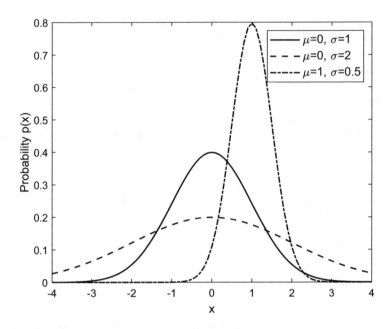

Fig. 11.2 Normal distributions for different values of μ and σ

$$f(x, a, b) = \frac{1}{b - a}$$

Here, a is the lower value for x and b is the upper value for x. All values between a and b have the same probability of occurrence and there is 0 probability of occurrence outside the range of a and b.

The relationship of parameters a and b to the mean and variance (the standard deviation squared) of the distribution is:

$$\mu = \frac{1}{2} (a + b)$$

$$\sigma^2 = \frac{1}{12} (b - a)^2$$

11.3.3 Lognormal

Not every probability distribution follows a normal or nearly normal distribution. The lognormal distribution represents a variable where the logarithm of the random

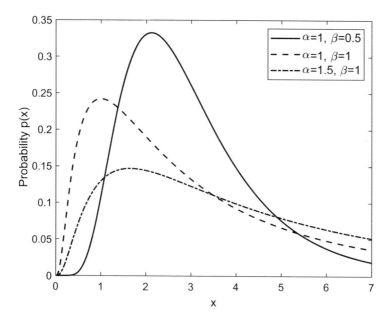

Fig. 11.3 Lognormal distributions for different values of α and β

variable is randomly distributed. The lognormal distribution probability density function is:

$$f(x, \alpha, \beta) = \frac{1}{x\beta\sqrt{2\pi}}\, e^{-\frac{(\ln x - \alpha)^2}{2\beta^2}}$$

The lognormal distribution represents a continuous variable which is always positive. The parameter α can take any value while β must be positive, similar to μ and σ from the normal distribution. There is no possibility that the variable has a value of zero, although there is a small probability that the value could be close to zero. The lognormal variable can represent a variety of real-world applications, including particle size distributions, size of living organisms, and personal income (Fig. 11.3).

The mean and variance of the lognormal distribution is given by the following expressions:

$$\mu = e^{\left(\alpha + \frac{\beta^2}{2}\right)}$$

$$\sigma^2 = \left(e^{\beta^2} - 1\right) e^{(2\alpha + \sigma^2)}$$

11.3.4   Weibull

The Weibull distribution can be used to model failure rate over time. The probability distribution function is given as:

$$f(x,\ \lambda,\ k) = \frac{k}{\lambda}\left(\frac{x}{\lambda}\right)^{k-1} e^{-(x/\lambda)^k}$$

Here, x is assumed to be greater than 0. Parameter λ is commonly referred to as the scale value and parameter k is called the shape factor. Both λ and k must be positive. Assuming that the value of x is the time-to-failure starting a $x = 0$, this distribution depends greatly on the value of parameter k. When $k < 1$ the failure rate decreases over time as defective items are removed from the population. When $k = 1$, the distribution represents a population with a constant failure rate. When $k > 1$, the Weibull distribution captures an aging process where older samples have a higher failure rate. Note that there are parameter values that give a nonzero probability for a value of 0. In terms of failure rate, this means a significant portion of manufactured items were defective at the time of their creation.

The mean and variance of the Weibull distribution is given by the following expressions involving the Gamma function Γ and the distribution parameters λ and k (Fig. 11.4).

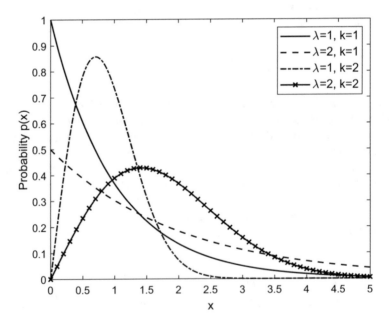

Fig. 11.4 Weibull distributions for different values of λ and k

$$\mu = \lambda \Gamma \left(1 + \frac{1}{k} \right)$$

$$\sigma^2 = \lambda^2 \Gamma \left(1 + \frac{2}{k} \right) - \left(\lambda \Gamma \left(1 + \frac{1}{k} \right) \right)^2$$

The Gamma function is defined as:

$$\Gamma(z) = \int_0^\infty x^{z-1} e^{-x} dx$$

which can be evaluated easily in MATLAB with the gamma function.

11.3.5 Beta

The Beta distribution probability distribution function is given as:

$$f(x, \alpha, \beta) = \frac{\Gamma(\alpha + \beta)}{\Gamma(\alpha)\Gamma(\beta)} x^{\alpha-1} (1 - x)^{\beta-1}$$

Here, x is assumed to be in the range 0 to 1. Parameters α and β must be positive values. These values have significant influence on the shape of the resulting distribution. The Beta distribution could be used to represent values that are constrained to values between 0 and 1 such as mole fractions (Fig. 11.5).

The mean and variance of the Beta distribution is given by the following expressions involving the gamma function Γ and the distribution parameters α and β.

$$\mu = \frac{\alpha}{\alpha + \beta}$$

$$\sigma^2 = \frac{\alpha\beta}{(\alpha + \beta)^2 (\alpha + \beta + 1)}$$

11.3.6 Student's t

In certain cases, only a limited number of sampled data is available. In other cases, mean values of two samples must be compared using only a few measurements. The distribution depends on the degrees of freedom, v and is given by Fig. 11.6:

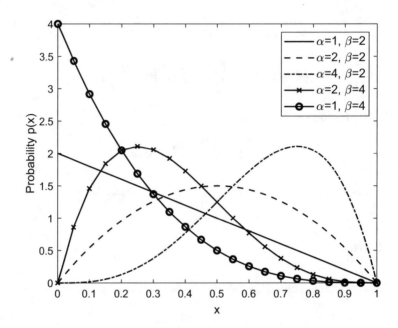

Fig. 11.5 Beta distributions for different values of α and β

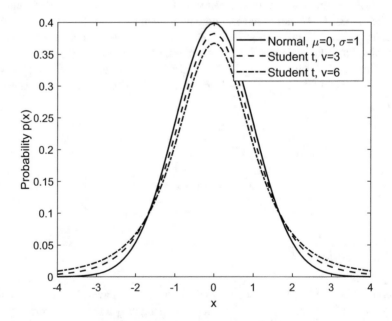

Fig. 11.6 Student t distribution for different values of v compared to a normal distribution with $\mu = 0$ and $\sigma = 1$

$$f(x) = \frac{\Gamma\left(\frac{1}{2}(v+1)\right)}{\sqrt{v\pi}\ \Gamma(v/2)\left(1+\frac{x^2}{v}\right)^{v+1}}$$

This distribution is often used for hypothesis testing to determine if there are differences between two groups of samples. This distribution gained notoriety after it was published under the pseudonym Student. William Gosset, at the request of employer the Guinness Brewery, published his work on small sample comparison anonymously (Fig. 11.6).

The Student t distribution is very similar to the normal distribution with $\mu = 0$ and $\sigma = 1$. The most noticeable difference lies in the larger tails of the Student t distribution. For large values of v, the Student t distribution and the normal distribution are almost indistinguishable. Taking n samples, the degrees of freedom v is defined as $v = n - 1$. The distribution represents the uncertainty on determining the sample mean. The mean and variance of the Student t distribution are:

$$\mu = 0$$

$$\sigma^2 = \frac{v}{v-2}$$

11.3.7 Discrete Uniform

The discrete uniform distribution considers discrete possibilities with equal probability of occurrence. Consider rolling an honest die, every number on the die should have equal possibility of occurrence. For a discrete uniform distribution over the inclusive interval $[a,b]$ the distribution is defined as:

$$f(x, a, b) = \frac{1}{b-a+1}$$

Now, the random variable x only takes on discrete values so the distribution function can be called the discrete density function or probability mass function. The values for a and b relate to the mean and variance by the following expressions:

$$\mu = \frac{a+b}{2}$$

$$\sigma^2 = \frac{(b-a+1)^2 - 1}{12}$$

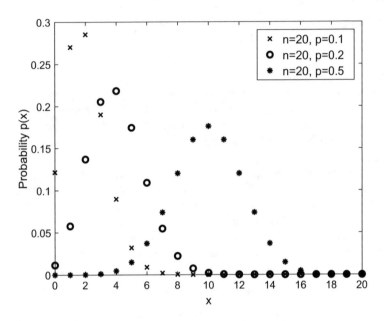

Fig. 11.7 Binomial distributions for different values of p with $n = 20$

11.3.8 *Binomial*

The binomial distribution is another discrete distribution. Instead of considering a single event, it represents n different events (Fig. 11.7). One may think of the binomial distribution as a true/false test with n questions where you have equal probability p for each question. The discrete density function is given as:

$$f(x, n, p) = \frac{n!}{k!\,(n - k)!} p^x \,(1 - p)^{n-x}$$

Here, n is a positive integer and p is a probability value in the range $(0, 1)$. When $p = 0.5$ the resulting distribution resembles a uniform distribution "bell" curve limited to values between 0 and n. The mean and variance relate to n and p as:

$$\mu = np$$

$$\sigma^2 = np\,(1 - p)$$

11.3.9 *Poisson*

The Poisson distribution describes the probability of a given number of events occurring with a specified average rate λ if the events occur independently. The discrete density function is given by:

$$f(x, \lambda) = \frac{\lambda^x e^{-\lambda}}{x!}$$

A well-known example of a Poisson distribution is the number of pieces of mail received each day when the average number of pieces of mail is fixed. Each letter arrives independently of the others. Some days no letters will arrive. The single parameter λ is the average rate value. The mean and variance are both equal to λ in the Poisson distribution (Fig. 11.8).

$$\mu = \lambda$$

$$\sigma^2 = \lambda$$

Table 11.1 shows a summary of different probability distribution functions and some of their distinguishing characteristics.

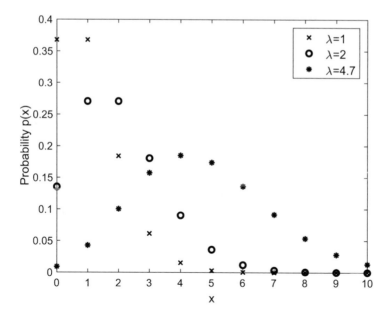

Fig. 11.8 Poisson distributions for different values of λ

Table 11.1 Comparison of distributions

Distribution	General type	Variable range	PDF/DDF
Normal	Continuous	$(-\infty, \infty)$	$\frac{1}{\sigma\sqrt{2\pi}}\,e^{-\frac{(x-\mu)^2}{2\sigma^2}}$
Continuous uniform	Continuous	$[a, b]$	$\frac{1}{b-a}$
Lognormal	Continuous	$(0, \infty)$	$\frac{1}{x\beta\sqrt{2\pi}}\,e^{-\frac{(\ln x-\alpha)^2}{2\beta^2}}$
Weibull	Continuous	$[0, \infty)$	$\frac{k}{\lambda}\left(\frac{x}{\lambda}\right)^{k-1}e^{-(x/\lambda)^k}$
Beta	Continuous	$(0, 1)$	$\frac{\Gamma(\alpha+\beta)}{\Gamma(\alpha)\Gamma(\beta)}x^{\alpha-1}(1-x)^{\beta-1}$
Student t	Continuous	$(-\infty, \infty)$	$\dfrac{\Gamma\left(\frac{1}{2}(v+1)\right)}{\sqrt{v\pi}\,\Gamma(v/2)\left(1+\frac{x^2}{v}\right)^{v+1}}$
Discrete uniform	Discrete	$[a, b]$	$\frac{1}{b-a+1}$
Binomial	Discrete	$[0, n]$	$\frac{n!}{k!(n-k)!}p^x(1-p)^{n-x}$
Poisson	Discrete	$[0, \infty]$	$\frac{\lambda^x e^{-\lambda}}{x!}$

11.4   Calculation of Basic Statistics

Given a set of N different measurement samples from a device or process representing a random variable, various statistics can be calculated to provide some insight into describing the random measurement. In many cases, the distribution parameters can be estimated using just the mean and process variance. Note that every distribution presented in Sect. 11.3 may be defined by the mean and standard deviation.

When sampling a process, care should be taken to ensure that the samples are truly independent and not biased. The sample should represent a true random sample from the possible measurements. For example, in a sheet-forming process such as paper making product samples should not be only taken from one edge of the sheet, rather samples should be taken at multiple points across the sheet. Other examples of potential sampling bias include:

- Political polling based only on people answering land line phone calls
- Paid clinical medical safety trials targeting economically disadvantaged
- Assumption that helmets in war lead to head injuries; dead never returned

These examples show that sampling bias can occur if the whole population of interest is not sampled in an unbiased manner.

11.4.1 Mean

The mean is just the average. The average of a sample of N values is readily calculated as:

$$\bar{x} = \frac{1}{N} \sum_{i=1}^{N} x_i$$

The sample mean $\bar{x}$ is the value determined from data while the mean μ is the true value of the distribution.

11.4.2 Median

The median is the middle value of the data. To determine the median value, first sort the data in ascending or descending numerical order. For data that has an odd value of N, the median is the middle value. For an even value of N, the median is the average of the two middle data values.

11.4.3 Variance

The variance describes the spread or variation found in a sample of data.

$$\sigma^2 = \frac{\sum_{i-1}^{N} (x_i - \mu)^2}{N - 1}$$

In cases where the entire population is measured, the variance equation is divided by N instead of $N - 1$. In most engineering cases, you will be taking a sample of measurements for a value, or a representative sample from a production run. The entire population is rarely measured.

The sample variance is sometimes distinguished from the true variance by assigning the sample variance as s^2 to differentiate from σ^2. Some authors will also use Var (X) to represent the variance of a random variable X.

11.4.4 Standard Deviation

The standard deviation is related to the variance. The standard deviation is a very common statistic used to indicate the spread or variation in a sample population.

Table 11.2 Comparison of
Normal and Student t
distributions

	Normal	t, $v = 3$	t, $v = 6$	t, $v = 25$
$\pm 1\sigma$	68.27%	60.90%	64.41%	67.31%
$\pm 2\sigma$	95.45%	86.07%	90.76%	94.35%
$\pm 3\sigma$	99.73%	94.23%	97.60%	99.40%
95%	$\pm 1.960\sigma$	$\pm 3.182\sigma$	$\pm 2.447\sigma$	$\pm 2.060\sigma$
99%	$\pm 2.578\sigma$	$\pm 5.841\sigma$	$\pm 3.707\sigma$	$\pm 2.787\sigma$
99.9%	$\pm 3.291\sigma$	$\pm 12.92\sigma$	$\pm 5.958\sigma$	$\pm 3.725\sigma$

$$\sigma = \sqrt{\sigma^2} = \sqrt{\frac{\Sigma_{i-1}^{N} (x_i - \mu)^2}{N - 1}}$$

Assuming a population follows a normal distribution, approximately 68% of any
sample measurement values will fall between $\mu \pm \sigma$. Similarly, 95% of samples
from a normal distribution will be between $\mu \pm 1.96\sigma$ and 99% will fall between
$\mu \pm 2.32\sigma$.

When only a limited number of experimental samples may be available, the
Student t distribution may be used to determine confidence intervals on estimates.
Note from Table 11.2 that the Student t and normal distributions are very similar for
larger sample sizes, while uncertainty is much greater for low numbers of samples
in the Student t distribution.

11.4.5   Standard Error

In some instances, the Standard Error (SE) is used to describe the variance of a
population:

$$SE = \frac{\sigma}{\sqrt{N}} = \frac{\sqrt{\Sigma_{i-1}^{N} (x_i - \mu)^2}}{\sqrt{N - 1}\sqrt{N}}$$

Note that when reporting any statistics, the sample size N should also be indicated.

The Standard Error is used to determine confidence intervals for a value. Two-
sided confidence intervals can be determined based on the desired confidence level
following:

$$CI = \bar{x} \pm z^* \frac{\sigma}{\sqrt{n}}$$

The z^* values are determined based on the assumption for the underlying distribution, as shown in the bottom section of Table 11.2. Assuming a normal distribution
for a variable, a sample mean $\bar{x}$, and n data points, the 95% confidence interval for
variable is:

$$-1.960\frac{\sigma}{\sqrt{n}} \leq \bar{x} \leq 1.960\frac{\sigma}{\sqrt{n}}$$

Note that this does not mean that 95% of the data will fall between $\pm 1.960\sigma/\sqrt{n}$. This confidence interval actually implies that given the data used to determine the sample mean $\bar{x}$ and σ, there is 95% confidence that the true mean μ is captured in the interval. Repeatedly running the same experiment 100 times, one would expect five of the experiments to return confidence intervals which did not include the true mean.

Example 11.1 (Statistics Calculation) Given the following data:

$$D = [4\ 3\ 7\ 2\ 1\ 2\ 4]$$

The average μ is found:

$$\mu = \frac{1}{N}\sum_{i=1}^{N} x_i = \frac{1}{6}(4 + 3 + 7 + 2 + 1 + 2 + 4) = 3.286$$

The median is found by ordering the data:

$$D = [1\ 2\ 2\ 3\ 4\ 4\ 7]$$

so the median is a value of 3, the middle value. The variance is found as:=

$$\sigma^2 = \frac{\sum_{i-1}^{N}(x_i - \mu)^2}{N - 1}$$

$$= \frac{\begin{array}{c}(4 - 3.286)^2 + (3 - 3.286)^2 + (7 - 3.286)^2 + (2 - 3.286)^2 \\ + (1 - 3.286)^2 + (2 - 3.286)^2 + (4 - 3.286)^2\end{array}}{7 - 1}$$

$$= 3.905$$

The standard deviation σ is simply $\sqrt{3.905} = 1.976$. The Standard Error is

$$SE = \frac{\sigma}{\sqrt{N}} = \frac{1.976}{\sqrt{7}} = 0.747$$

11.5   Central Limit Theorem

The central limit theorem maintains that when any random variable is properly sampled, the mean of the samples will approach a normal distribution for a sufficiently large number of samples. This does not mean that "Everything is Gaussian" or that only a handful of samples are required.

Consider a Poisson process with $\lambda = 3$ as seen in Fig. 11.9. The Poisson process is a discrete process where the parameter λ defines both the mean and standard deviation, $\lambda = \mu = \sigma$. Given only a few samples, examining the histogram gives little insight into the process. For $N = 2000$ random samples, the histogram approaches the theoretically predicted PDF shown with "x" values in the top right graph of Fig. 11.9.

Given the Poisson process with $\lambda = 3$, one may take N samples from the random distribution and compute the sample mean. Repeating this random sampling 200 times with different numbers of samples in each histogram is shown in the bottom row of Fig. 11.9. In all three cases, the mean of the sample means approaches the true value of 3. For larger values of N, the variation in the sample mean calculation is reduced.

The underlying message is that the random variable may not be normally distributed but when you sample that variable N times and determine the sample mean, the resulting sample mean will itself be a random value following a Gaussian distribution. Using the central limit theorem, experimental sample means can be

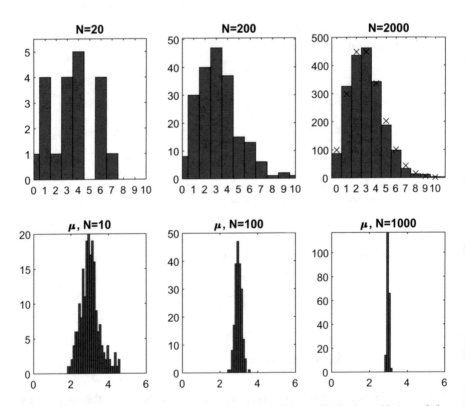

Fig. 11.9 Top row: Histograms for randomly evaluated Poisson distribution with $\lambda = 3$ for different numbers of samples, $N = 20$, $N = 200$, and $N = 2000$. Bottom row: histograms for estimating the Poisson distribution mean value. Each was computed 200 different times using 10 samples, 100 samples, and 1000 samples. The sample mean exhibits a normal distribution with the mean of the means approaching 3. Using more random samples results in lower variability of the mean

determined and compared between two populations. This leads to hypothesis testing for determining with what confidence two samples may be said to be from different sample populations.

11.6 Hypothesis Testing

The most useful application of statistics arises from the need to determine what can be said about two different populations. Are the two samples from the same population? Or are the two samples actually different? What level of confidence do you have that the samples are different? This type of hypothesis testing for comparing effects is used in many applications, from drug testing to chemical processing.

Consider two sets of data, A and B. Each set of data represents different experimental conditions. It is desired to determine if the two sets of data are actually different. Did changing the conditions from experiment A to experiment B make a difference? First, one must pose the null hypothesis.

> **Null Hypothesis**—The null hypothesis assumes that there is no significance difference between two experiments.

The goal of hypothesis testing is to compare data sets to see if the null hypothesis can be invalidated. Basically, is there enough information to say that the null is false and therefore data sets for A and B are different. Further, statistics can give an estimate of the confidence in the assessment.

When performing a hypothesis test, the significance level is specified.

> **Significance Level**—The significance level α is used to determine the bound on the test statistic which represent the rejection of the null hypothesis.

One may use 95% confidence limits to determine if two experiments are different. This means setting the significance level $\alpha = 0.05$. Note that there are two-sided and one-sided hypothesis tests. A two-sided test would be useful for considering is $\mu_A = \mu_B$. Alternatively, a one-sided hypothesis test would be useful for considering a question like is $\mu_A > \mu_B$ or possibly $\mu_A < \mu_B + 2$ when a large difference is desired.

The hypothesis test gives a single limit to either reject or accept the null hypothesis. However, examining the p value can give even more useful insight to the level of certainty.

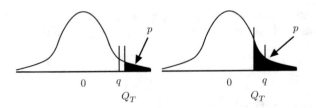

Fig. 11.10 Two different cases for one-sided hypothesis testing. If the test statistic Q_T exceeds the critical value q, the null hypothesis can be rejected. The p value represents the probability that the sample is actually a rare event and the null should not be rejected. When $Q_T < q$, the p value exceeds the significance level α and there is a strong possibility that the sample is not interesting so that the null hypothesis is not rejected

---

> **p Value**—The p value represents the probability of obtaining more extreme values given the provided data.

For a data set making a one-sided comparison such as $\mu_A > \mu_B$, the significance level could be set to $\alpha = 0.05$. This will result in a t-test value. If the test statistic for μ_A is greater than the determined test value, the null hypothesis is rejected and you can say with 95% confidence that $\mu_A > \mu_B$. However, if μ_A is very close to the test limit, there is a possibility ($\alpha = 0.05$) that the result is in error. The p value in this case would be very close to 0.05. If the p value is smaller, this means you have additional confidence that the data represents confidence in rejection of the null hypothesis (Fig. 11.10).

When comparing two sets of data, the critical value q is determined based on the prescribed significance level then the test statistic Q_T is determined based on the data. The null hypothesis for a one-sided comparison is based on:

$$Q_T \geq q$$

For a two-sided hypothesis Q_T must be less than or greater than the limit to reject the null hypothesis. There are different ways to determine t-test values for Q_T and q depending on problem assumptions. Three common options include:

- Assuming equal variances in samples
- Assuming unequal variances in samples
- Assuming paired data

For paired samples, each individual is measured twice for the comparison, as in testing of weight-loss programs. Many different statistical software packages are available to quickly evaluate t-test values, including MATLAB and Excel. Table 11.3 shows data for statistical comparison of $A1 > B$ and $A2 > B$ with $\alpha = 0.05$. Both sets of data show that $A > B$ since the p values are $>= 0$. Data set $A1$ is very close to the border for statistical significance since the p value is very close to 0.05, the prescribed α value.

Table 11.3 Two t-test comparisons of $\mu_A > \mu_B$ for two A data sets assuming unequal variances in the populations

	μ_B	μ_{A1}	μ_{A2}
	4.1	5.1	5.1
	4.3	4.8	4.8
	4.5	4.6	4.7
	4.8	4.3	4.5
	4.2	5.2	5.5
	4.33	4.3	4.6
$\bar{x}$	4.37	4.72	4.87
σ^2	0.062	0.150	0.139
t-test value		1.837	2.71
t-test limit		1.833	1.833
p value		0.050	0.012

11.6.1 Type I and Type II Errors

Statistical hypothesis testing includes the possibility of making mistakes. There are two types of error, false positive and false negatives. These types of error are referred to as Type I and Type II error.

> **Type I Error**—Type I Error erroneously rejects the null hypothesis.

> **Type I Error**—Type II Error erroneously fails to reject the null hypothesis.

Type I error rejects the null hypothesis incorrectly. Some examples of Type I error include:

- Incorrectly quarantining a healthy individual (false positive)
- Convicting an innocent man
- Rejecting a good shipment of incoming material

Type I error is also known as α risk or producer risk. When a producer sends out a shipment of goods, the shipment is usually sampled and quality tested upon arrival. If the testing results in an erroneous outcome, the shipment would be rejected (Fig. 11.11).

Type II error fails to reject the null hypothesis. Some examples of Type II error include:

- Incorrectly allowing a contagious individual to roam (false negative)
- Setting free a criminal
- Consumer accepting a bad shipment of goods

Type II error is also known as β risk or consumer risk. If testing a shipment fails to recognize that the shipment is low-quality, the consumer will have accepted a

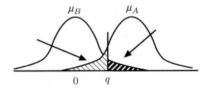

Fig. 11.11 Comparison of α and β risk for considering $\mu_A > \mu_B$. When the sample is actually from group B, there is possibility α that the null hypothesis ($\mu_A > \mu_B$) is erroneously rejected (Type I). If the sample is actually from group A, there is probability of β that the null hypothesis is erroneously not rejected (Type II)

shipment of low-quality material. The α value can be adjusted. However, this value directly impacts the β value. A stronger hypothesis test will result in two populations with wide separation in the sample means, allowing for very little population overlap and chance for either Type I or Type II error.

11.7   Applied Methods

Industrial applications for production often use statistical tools to help ensure consistent product quality. These tools often go by a variety of names, including:

- Six Sigma
- Lean Manufacturing
- Lean Six Sigma

These methods often include a variety of techniques. Many of the methods use statistical methods that can be applied without deep understanding of the underlying methodologies. Brief descriptions of some of the industrial methods are provided here.

Six Sigma methods usually focus on a process producing quality products such that the product average including $\pm 6\sigma$ of uncertainty meets quality standards, even if the mean shifts by $\pm 1.5\sigma$. This implies that 99.999% of the product meets standards even when the process mean shifts by a significant amount.

Control Limits—Upper and lower limits for a measured value in a production process representing the range of acceptable product quality.

DPMO—Defects Per Million Opportunities. A measure of how often a process is producing products that do not meet product quality specifications.

Six Sigma—Manufacturing a product with product quality control limits set to $\pm 3\sigma$ of the process mean, allowing for $\pm 1.5\sigma$ shift in the mean up or down.

Lean Manufacturing—Quality manufacturing strategy involving customer value, analysis of the value stream, product flow through the manufacturing process, introducing steps to help pull a product through the process, and ultimately working toward manufacturing perfection.

DMAIC—Design, Measure, Analyze, Improve, Control. Cyclic strategy used for approaching problems in manufacturing to improve product quality.

DMADV—Define, Measure, Analyze, Design, Verify. Cyclic strategy used for product design.

Check Sheets—Collection of manufacturing data at the point where the data is generated.

Control Charts—Plotting product quality values over time to help detect when a production process is trending away from normal.

SPC—Statistical Process Control. Use of control charts to monitor product quality.

Nelson Rules—Rules for analysis of control charts to determine when to initiate action to get product quality back into specification.

Flow Chart—A graphical representation of a production process indicating discrete steps.

Pareto Chart—A chart of product quality problems listed in decreasing order of frequency.

Fishbone Diagram—Graphical representation of causes leading to product defects. Sometimes referred to as Ishikawa diagram or cause-and-effect diagrams.

Scatter Diagrams—XY plot of two variables to help find trends in data.

PICK Chart—Possible, Implement, Change, Kill. Categorization of changes based on high/low payoff and high/low cost to implement.

PDCA—Plan, Do, Check, Adjust. Cyclic strategy to improve quality.

TOC—Theory of Constraints. Identify process constraints, exploit constraints, remove constraints, repeat as needed to improve quality.

OEE—Overall Equipment Effectiveness. A measure of process availability, performance, and quality.

FMEA—Failure Mode Effects Analysis.

Cpk—Process capability index. $\min\left\{\frac{U_L-\mu}{3\sigma}, \frac{\mu-L_L}{3\sigma}\right\}$ Represents how much a process can change without violating upper and lower productions limits, U_L and L_L.

11.8   Monte Carlo Methods

Given a probability distribution, it is often difficult to mathematically predict the expected result to answer useful questions. While it may be possible to derive the answer mathematically, it may be much simpler to use a simulation-based approach. Due to the vast speed of computers, simulation-based approaches can often be reasonable to use in calculation of probabilistic quantities.

Example. Given three six-sided dice randomly rolled and added together, what is the total probability of rolling 16, 17, or 18? There are $6 \times 6 \times 6 = 216$ different possible combinations of three six-sided dice. To roll a 18, 17, or 16 there are ten different combinations:

18 :{6, 6, 6}

17 :{5, 6, 6} or {6, 5, 6} or {6, 6, 5}

16 :{4, 6, 6} or {6, 4, 6} or {6, 6, 4} or {5, 5, 6} or {5, 6, 5} or {6, 5, 5}

So the exact probability of 16, 17, or 18 occurring is 10/216 or 0.0463, about 4.63% of the time you roll three dice together.

To approximate this using a Monte Carlo approach, the following MATLAB code could be used:

```
N=100000;
sum(sum(randi(6,3,N))>=16)/N
```

This code makes a matrix with 3 rows and N columns containing integers 1 to 6 sampled from the uniform distribution. The total number of instances with the sum$\geq$ 16 is determined. As N, the number of simulations increases, the value converges to the true exact value. The following figure shows 20 estimated values using a Monte Carlo simulation with N runs as a function of N. As you can see, as N increases, the variation in the estimated value decreases and eventually converges to the true value (Fig. 11.12).

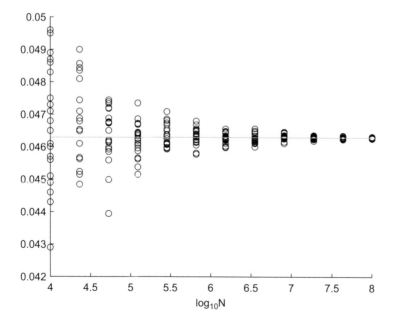

Fig. 11.12 Comparison of 20 simulations of size N for probability estimation of rolling a sum of 16, 17, or 18 with three six-sided dice

Monte Carlo methods can give estimates for true values but they are not exact solutions. In some cases, the exact solution can be found using probability calculations. Detailed discussion of the required combination and permutation calculations is beyond the scope of this chapter. In many cases, exact determination of the value in question can be quite difficult or even impossible. Monte Carlo approaches can be used to estimate the value using a simulation approach.

There are two significant disadvantages to Monte Carlo approaches. The first disadvantage relates to rare events. A simulation-based approach to probability calculations can fail to capture extremely rare events. If a significant event occurs with a very low probability, it may never occur in the many simulations run. Consider simulating winning the lottery with chances around 1 in 100,000,000. Running the random test 1,000,000 times would give a very small chance of a single successful event.

The second disadvantage is that a large number of simulations may be required to get accurate estimates for a value. In the previous example, a specific single value is being estimated. Increasing the number of simulation runs helps reduce uncertainty in the estimate. In other situations, Monte Carlo methods may be used to see how uncertainty affects a process model. The result is not a scalar value, but rather a distribution of values representing how the assume uncertainty propagates through a model. If the simulation requires significant calculation for each simulation run, the resulting computation time may be prohibitive.

Problems

11.1 Given data for concentration measurements:

$$C = \begin{bmatrix} 45 & 77 & 81 & 95 & 52 & 60 & 53 \end{bmatrix}$$

(a) Find the mean
(b) Find the median
(c) Find the variance
(d) Find the standard deviation
(e) Find the standard error

11.2 Given the measurements:

$$x = \begin{bmatrix} 1.1 & 2.1 & 1.8 & 0.91 & 1.2 & 1.7 & 0.98 \end{bmatrix}$$

(a) Find $\bar{x}$
(b) Find the median of x
(c) Find σ
(d) Find SE

11.3 Given data for concentration measurements:

$$T = \begin{bmatrix} 215 & 217 & 222 & 209 & 211 & 223 & 219 \end{bmatrix}$$

(a) Find the mean
(b) Find the median
(c) Find the variance
(d) Find the standard deviation
(e) Find the standard error

11.4 Given data for concentration measurements:

$$W = \begin{bmatrix} 0.7 & 0.6 & 0.7 & 0.8 & 0.4 & 0.9 & 0.5 \end{bmatrix}$$

(a) Find the mean
(b) Find the median
(c) Find the variance
(d) Find the standard deviation
(e) Find the standard error

11.5 Given data for concentration measurements:

$$D = \begin{bmatrix} 3 & -1 & 4 & 0 & 5 & -2 & 2 \end{bmatrix}$$

(a) Find the mean
(b) Find the median
(c) Find the variance
(d) Find the standard deviation
(e) Find the standard error

11.6 Given data that fits a **uniform** probability distribution between 0 and 5:

(a) Sketch the probability distribution function
(b) Determine the average value for this distribution
(c) Provide one example of a uniform probability distribution in nature

Further Reading

1. Council for Six Sigma Certification. (2018). *Six Sigma: A complete step-by-step guide*. The Council for Six Sigma Certification.

2. Devore, J.L., Farnum, N.R., & Doi, J.A. (2013). *Applied statistics for engineers and scientists* (3rd ed.). Cengage Learning.
3. Montgomery, D.C., & Runger, G.C. (2013). *Applied statistics and probability for engineers* (6th ed.). Wiley.
4. Navidi, W. (2019). *Statistics for engineers and scientists* (5th ed.). McGraw-Hill Education.
5. Ogunnaike, B.A. (2009). *Random phenomena: Fundamentals of probability and statistics for engineers* (1st ed.). CRC Press.
6. Ries, J. (2018). *Lean mastery collection: 8 books in 1 - Lean Six Sigma, Lean Startup, Lean Enterprise, Lean Analytics, Agile Project Management, Kanban, Scrum, Kaizen).* Independently published.

Chapter 12
Linear Modeling

12.1 Linear Interpolation

In many cases, scientific data may be provided in tabular form. For example, the tensile strength of a polymer may depend on its temperature value. Typically, these values are determined from repeated carefully run laboratory experiments, ideally replicated at multiple locations under identical conditions.

This data is useful, but not every case is considered. Examine the example polymer data shown in Table 12.1. What if the product design required that the polymer, what if the engineering design required that the polymer be used at an operating temperature of $125\,°C$? Rather than attempting to run the original laboratory experiment at a temperature of $125\,°C$, you could use the average value of the two closest data values, 768 MPa and 430 MPa, resulting in a value of $1198/2 = 599$ MPa. This may not be the exact value, but it could be an adequate initial approximation. What if the design temperature value was not exactly between the two temperature values? Say the design temperature is $142.7\,°C$?

To estimate the value at an intermediate value, the concept of linear interpolation can be used. Given two data points, a value on the line between the two points can be estimated using the formula:

$$y = f(x) \approx f(x_0) + (x - x_0)\,\frac{f(x_1) - f(x_0)}{x_1 - x_0}$$

Here, it is assumed that the estimated value is a function of x, $y = f(x)$. It is also assumed that $x_0 < x_1$, x_0 is the "left" value on the interval and x_1 is the "right" value. This formula is just the formula of a line in the form of $y = b + mx$, with a slope of:

$$m = \frac{f(x_1) - f(x_0)}{x_1 - x_0}$$

© Springer Nature Switzerland AG 2022
E. Gatzke, *Introduction to Modeling and Numerical Methods for Biomedical and Chemical Engineers*, https://doi.org/10.1007/978-3-030-76449-4_12

Table 12.1 Example data
table

Tensile strength (MPa)	Temperature (°C)
876	50
768	100
430	150

Remember that you must identify the dependent value that *depends* on the independent value. In the polymer tensile strength example, when given a temperature value, the tensile strength can be interpolated from the data. Similarly, given a desired tensile strength value, the requisite temperature could be determined using the same formulation under the assumption that the data is monotonically increasing or decreasing.

Example 12.1 (Interpolation) The viscosity of a fluid measures its resistance to deformation. It can be measured in unit of centiPoise, cP. Given the tabular data below, determine the viscosity of the fluid at $117\,°C$ using linear interpolation:

Viscosity (cP)	°C
0.95	100
0.92	120
0.86	140
0.79	160
0.70	180
0.44	200

The temperature of $117\,°C$ is between the table values of $100\,°C$ and $120\,°C$. This means that the resulting value will be between 0.95 cP and 0.92 cP. Since $117\,°C$ is close to $120\,°C$, you should expect the interpolated value to be close to 0.92 cP. The formula for interpolation is:

$$y = f(x) \approx f(x_0) + (x - x_0)\,\frac{f(x_1) - f(x_0)}{x_1 - x_0}$$

Using the values from the table with $x_0 = 100°$ and $x_1 = 120°$ and the value $x = 117°$, the equation becomes:

$$y \approx 0.95 + (117 - 100)\,\frac{0.92 - 0.95}{120 - 100} = 0.9245$$

This seems reasonable, as 117 is close to 120, so the values should be quite similar. Additionally, note that the slope value is negative in this case, as the viscosity values decrease with temperature:

$$m = \frac{f(x_1) - f(x_0)}{x_1 - x_0} = \frac{0.92 - 0.95}{120 - 100} = -0.0015$$

The final answer should also be reported with units: 0.9245 cP.

Linear interpolation can be very useful to estimate values between two points of data. However, the underlying assumption remains that the variable of interest will change linearly between the two points. Obviously, more advanced methods could be used by assuming that the function follows a nonlinear curve between the two points. A linear approximation should be accurate enough for most applications. If higher levels of accuracy are required for values in a highly nonlinear system, more data points may be required rather than attempting to determine values using linear interpolation.

12.2 Linear Extrapolation

In some cases, tabulated data may not include the range of interest. The desired value could be outside the reported range of values. Extrapolation can be used to approximate the required value. However, extrapolation of data values should always be done with great care. A real physical system could have significant changes outside the range of reported values. A phase change could occur outside the temperature range or a different physical phenomenon could start to dominate the determination of the measured value.

The same interpolation equation can be used for extrapolation:

$$y = f(x) \approx f(x_0) + (x - x_0) \, \frac{f(x_1) - f(x_0)}{x_1 - x_0}$$

The two closest data points can be used as x_0 and x_1. Again, the "left" data point is x_0 and the "right" data point is x_1.

Example 12.2 (Extrapolation) Consider again the viscosity of a fluid. Given the tabular data below, determine the viscosity of the fluid at 96 °C and 211 °C using linear extrapolation:

Viscosity (cP)	°C
0.95	100
0.92	120
0.86	140
0.79	160
0.70	180
0.44	200

The temperature of 96 °C is below the reported table values. The two closest values are 100 °C and 120 °C. Using the values from the table with $x_0 = 100$°C and $x_1 = 120$°C and the value $x = 96$°C, the equation becomes:

$$y \approx 0.95 + (96 - 100) \ \frac{0.92 - 0.95}{120 - 100} = 0.956$$

This value seems reasonable, as 96 is close to 100. The viscosity values are increasing for decreasing temperature values, so it is expected that the extrapolated value will exceed the value at $100\,°C$. Again, the final answer should also be reported with units: 0.956 cP.

To find the value at $211\,°C$, the same formula is used with data points for $180\,°C$ and $200\,°C$. The x value is now $211\,°C$, resulting in:

$$y \approx 0.70 + (211 - 180) \ \frac{0.44 - 0.70}{200 - 180} = 0.297$$

The value of 0.297 cP is reasonable. The negative slope value Means that viscosity values increase as the temperature decreases. The biggest concern is that a phase change occurs outside of the range of the table temperature values, making the extrapolation estimates invalid.

Again, be very careful when extrapolating values for data. Only use extrapolation as a last resort when a value must be estimated from existing data. Phase changes or fundamental changes in the underling physics can make extrapolation of data values completely invalid.

12.3   Linear Regression

Given a data for x and y, plotting the data often results in a linear relationship. Assuming that $y = f(x) = mx + b$, the constants for slope m and intercept b should be found to create a linear model. Rather than just pick two points to use in determining m and b, all of the available data can be used in a single calculation to find the best-fit parameter values. This process is called linear regression or data reconciliation. In some cases, the resulting linear model is called a trend line.

Model parameters that appear linearly can be formulated as a non-square solution to a linear algebraic system of equations. Assume that the scalar measured value y depends on a process parameter x. Assume that the model takes the form:

$$y = mx + b \tag{12.1}$$

Technically, you only need two data points to find m and b, the model parameters. Assuming that you have more than two data points, we often desire to determine the "best-fit" for the line. These parameters minimize the sum of the square of the model error. Consider an experiment with four data points:

$$y_1 = m\,x_1 + b$$

$$y_2 = m\,x_2 + b$$

$$y_3 = m\,x_3 + b$$

$$y_4 = m\,x_4 + b \qquad\qquad (12.2)$$

Here, values of y_1 to y_4 and x_1 to x_4 are known, but m and b are unknown values. This can be written as a set of equations:

$$\begin{bmatrix} y_1 \\ y_2 \\ y_3 \\ y_4 \end{bmatrix} = \begin{bmatrix} x_1 & 1 \\ x_2 & 1 \\ x_3 & 1 \\ x_4 & 1 \end{bmatrix} \begin{bmatrix} m \\ b \end{bmatrix}$$

This is a set of equations in the form of $Ax = b$ with a non-square A matrix. The vector b is made up of measurement values. The unknown vector x represents the unknown values for the slope and the intercept. The "best-fit" solution to this over-specified set of equations may be found using the pseudo-inverse of the matrix A:

$$x = \left(A^T A\right)^{-1} A^T b$$

Here, A^T is the transpose of the matrix A. The product $A^T A$ results in a square matrix, which potentially will have an inverse. This type of regression will minimize the error between the measurements y_1 to y_4 versus the model predictions. Specifically, this minimizes the sum of the errors squared. This is commonly known as minimizing the Sum Squared Error (SSE). The value $\left(A^T A\right)^{-1} A^T$ is called the Moore–Penrose pseudo-inverse.

12.3.1 Correlation Coefficient

A goodness of fit is often reported when performing a linear regression (making a trend line). This r^2 value is the correlation coefficient. If the r^2 value is 1.0, this means that the model is a perfect fit to the data; the model perfectly fits every single data point. Due to the measurement error, this should never happen in real life (unless you only have two data points). Usually, the r^2 value for a good model fit is around 0.99 or 0.95. Even adequate models may be found for r^2 values around 0.8 or lower.

12.4   Transformation to Linear Form

Most real systems are nonlinear. However, a simple linear transformation on the nonlinear equations can make the nonlinear relationship appear linear. This can be exploited to determine model values from data values. The linear relationship can be plotted, and a linear regression can lead to determination of parameter values.

A common mathematical method used by engineers is a transformation of an equation from a nonlinear form into a linear form. Consider the simple nonlinear equation defining how a concentration value $C(t)$ changes with time:

$$C(t) = ae^{-kt}$$

Given data values for concentration at various points in time, it is desired to find the numerical values for model parameters a and k. Of course, to find two values, one technically only needs two data points. However, due to noisy data collection, many data points should be used to find the best values. This equation can be transformed by taking the natural log on both sides:

$$\ln(C(t)) = \ln(a) - kt$$

This equation has a linear form, $y = mx + b$. The values for $C(t)$ can be easily used to calculate $\ln(C(t))$. These values represent y values in the transformed problem. The values for time t can be used as the x value without transformation. Plotting $\ln(C(t))$ vs. t should provide a linear graph with a slope m and an intercept b. From the transformed problem, the slope of the plot can be used to determine the unknown value k, since $m = -k$. The intercept of the linear plot can be used to determine a, since $b = \ln(a)$.

Example 12.3 (Transformation to Linear Form) Consider the example where a reaction rate value is assumed to be an exponential function of reactor temperature T:

$$r = Ae^{-B/T}$$

This equation is nonlinear when measured r values are plotted against temperature values T. Taking the natural log on both sides leads to the equation:

$$\ln(r) = \ln(A) - \frac{B}{T}$$

This relationship does not appear to be linear. However, treating $\ln(r)$ as y and the value $1/T$ as x makes the relationship linear. Plotting values for $\ln(r)$ vs. $1/T$ should result in a straight line. The slope corresponds to the value $-B$ and the intercept value will determine $\ln(A)$, from which the value of A can be found.

Example 12.4 Transform to linear form

A simple chemical process is assumed to conform to the following equation:

$$\frac{k_2 C(t) t - k_1 t - 1}{k_2} = 0$$

Here, the concentration $C(t)$ is a function of time t. The model depends on model parameters k_1 and k_2. Assume that you are provided concentration measurement values at points in time, $C(t)$, and values t.

This equation is not currently in a linear form. The two unknown values are k_1 and k_2. The equation must be in the linear form

$$y = mx + b$$

The equation can be rearranged a bit. First, divide the numerator terms by k_2:

$$\frac{k_2 C(t) t}{k_2} - \frac{k_1}{k_2} t - \frac{1}{k_2} = 0$$

$$C(t) t - \frac{k_1}{k_2} t - \frac{1}{k_2} = 0$$

Now, the equation terms can be divided by t.

$$\frac{C(t) t}{t} - \frac{k_1 t}{k_2 t} - \frac{1}{k_2} \frac{1}{t} = 0$$

$$C(t) - \frac{k_1}{k_2} - \frac{1}{k_2} \frac{1}{t} = 0$$

Next, bring two terms to the right hand side:

$$C(t) = \frac{1}{k_2} \frac{1}{t} + \frac{k_1}{k_2}$$

This is now linear, although it does not appear so at first. Since you are given numeric values for $C(t)$ and t, the data values could be used to plot a straight line. Let $y = C(t)$ and $x = \frac{1}{t}$ so that the linear form appears as:

$$y = \frac{1}{k_2} x + \frac{k_1}{k_2}$$

Now, it can be seen that plotting $y = C(t)$ vs. $x = \frac{1}{t}$ should give a straight line with slope $m = \frac{1}{k_2}$ and intercept $b = \frac{k_1}{k_2}$. Once the slope is found, the value for k_2 can be determined. Given that k_2 is known, the value for k_1 can be found from the intercept value.

12.5   Linearization

In some cases, algebraic rearrangement of a nonlinear function may not lead to a linear form. As seen in Chap. 8, a first-order Taylor series may be used to create a linear approximation at a single point. The resulting linear approximation should be 100% accurate at the point where the linearization was created. At other points away from that value, the linear model will be less accurate.

Consider a function $f(x)$, a scalar function of one variable. A Taylor series can be used to approximate the function near a point x^* as

$$f(x) \approx f(x)|_{x=x^*} + \frac{f(x)'|_{x=x^*}}{1!}(x-x^*)^1 + \frac{f(x)''|_{x=x^*}}{2!}(x-x^*)^2 + \frac{f(x)'''|_{x=x^*}}{3!}(x-x^*)^3 + \dots$$

The value of the function and its derivatives must be evaluated at $x = x^*$. The first-order approximation leads to a linear form by dropping higher order terms:

$$f(x) \approx f(x) \approx f(x)|_{x=x^*} + f(x)'\Big|_{x=x^*}(x-x^*)$$

This is the basis for Newton's method. The nonlinear function is approximated in the form of $y = mx + b$, where x can be solved as $x = \frac{y-b}{m}$. Linearization is used in other cases such as dynamic systems.

Example 12.5 (Linearization of a Dynamic System) Given the nonlinear dynamic model for a draining tank with a constant inlet flow rate:

$$\frac{dh}{dt}(t) = 2 - \sqrt{h(t)}$$

At steady-state, $\frac{dh}{dt} = 0$, so $h_{ss} = 4$. The nonlinear term $\sqrt{h(t)}$ can be approximated using a Taylor series as:

$$\sqrt{h(t)} \approx \sqrt{h_{ss}} + \frac{1}{2}(h_{ss})^{-1/2}(h(t) - h_{ss}) = 2 + \frac{1}{4}(h(t) - h_{ss})$$

Introducing the new variable $y(t) = h(t) - h_{ss}$ and recognizing that $\frac{dy}{dy} = \frac{dh}{dt}$ leads to:

$$\frac{dy}{dt}(t) = 2 - \left(2 + \frac{1}{4}y(t)\right) = -\frac{1}{4}y(t)$$

The solution to this equation is $y(t) = y_0 e^{-\frac{1}{4}t}$. We know that $y(t) = h(t) - h_{ss}$ so that if $h(t = 0) = 0$, then $y(t = 0) = -4$, and the approximate solution for filling the tank is:

$$h(t) = -4e^{-\frac{1}{4}t} + 4$$

Problems

12.1 Use linear interpolation or linear extrapolation on the following problems using the tabulated data. Show your work.

(a) Estimate y (44)
(b) Estimate y (98)
(c) Estimate y (109)
(d) Estimate y (5)
(e) Estimate y (135)

$y = f(x)$	x
98.1	20
73.2	40
49.8	60
36.9	80
11.2	100

12.2 Use linear interpolation or linear extrapolation on the following problems using the tabulated data. Show your work.

$y = f(x)$	x
−2.1	0
−1.5	1
−0.7	2
0.3	3
0.8	4

(a) Estimate y (0.4)
(b) Estimate y (2.3)
(c) Estimate y (2.8)
(d) Estimate y (−0.4)
(e) Estimate y (5)

12.3 Use linear interpolation or linear extrapolation on the following problems using the tabulated data. Show your work.

T (°C)	C_A $\left(\frac{mol}{L}\right)$
100	2.34
150	2.78
200	3.10
250	3.35
300	3.87

(a) Estimate the concentration at $T = 114\,°C$
(b) Estimate the concentration at $T = 211\,°C$
(c) Estimate the concentration at $T = 97\,°C$
(d) Estimate the concentration at $T = 321\,°C$
(e) Estimate the temperature required to achieve a concentration of 2.45 $\frac{mol}{L}$
(f) Estimate the temperature required to achieve a concentration of 3.15 $\frac{mol}{L}$
(g) Estimate the temperature required to achieve a concentration of 3.93 $\frac{mol}{L}$
(h) Estimate the temperature required to achieve a concentration of 2.30 $\frac{mol}{L}$

12.4 Use linear interpolation or linear extrapolation on the following problems using the tabulated data. Show your work.

F $\left(\frac{L}{s}\right)$	T (C)
10	114
20	103
30	92
40	85
50	76

(a) Estimate the temperature at a flow of 14.5 $\left(\frac{L}{s}\right)$
(b) Estimate the temperature at a flow of 41 $\left(\frac{L}{s}\right)$
(c) Estimate the temperature at a flow of 64 $\left(\frac{L}{s}\right)$
(d) Estimate the temperature at a flow of 4.1 $\left(\frac{L}{s}\right)$
(e) Estimate the flow at $T = 121\,°C$
(f) Estimate the flow at $T = 84\,°C$
(g) Estimate the flow at $T = 107\,°C$
(h) Estimate the flow at $T = 63\,°C$

Chapter 13
Forces and Moments

13.1 Vectors

Vectors in linear algebra are stacked numbers that arise from sets of linear equations, such as the vectors $\underline{x}$ and $\underline{b}$ in a set of equations $\underline{\underline{A}}\,\underline{x} = \underline{b}$. Vectors can also have a more physical meaning. The two types of vectors engineers commonly work with are velocity and force.

Vector—A vector has both direction and magnitude.

The velocity of an object is not the same as its speed. Velocity includes direction information, while speed is just the magnitude (size) of the velocity. Velocity is easy to visualize because everyone is familiar with moving objects. When you throw a ball or drive a car, the velocity determines how fast the object moves and where the objects move toward.

Force vectors may be a bit more difficult to visualize. The force on an object may come from Earth's gravity pulling it down toward the center of the earth. The force on an object could come from another object pushing or pulling on it. Magnetic fields can exert forces on ferrous materials. These forces can work to move objects. For an object with constant mass m under a constant force F, the governing equation is $F = ma$, where a is acceleration, the rate of change of velocity.

This chapter considers relatively simple problems involving objects that are not moving. Only simple constant forces are exerted on the objects. Later classes may consider much more complicated problems involving changing velocity, changing mass, or changing forces.

© Springer Nature Switzerland AG 2022
E. Gatzke, *Introduction to Modeling and Numerical Methods for Biomedical and Chemical Engineers*, https://doi.org/10.1007/978-3-030-76449-4_13

13.1.1 Force Vector Components

To describe a vector requires that you define the size of the vector and the direction of the vector. To start defining a vector in two or three dimensions, one must first understand the unit vectors in the x, y, and z directions. These unit vectors provide a basis or reference for defining force vectors.

Unit Vectors—The unit vector **i** is magnitude 1 and points in the positive x direction. The unit vector **j** is magnitude 1 and points in the positive y direction. The unit vector **k** is magnitude 1 and points in the positive z direction.

For simplicity, we will only consider two-dimensional vectors. The overall force vector can be expressed as a sum of the x component and the y component. Assuming that the positive x direction is defined by a unit vector **i** and that the positive y direction is defined by a unit vector **j**, the force F can be expressed as:

$$F = F_x\mathbf{i} + F_y\mathbf{j}$$

Here, F_x is the magnitude of F in the x direction. F_y is the magnitude of F in the y direction. The components F_x and F_y of a vector can be positive, negative, or 0. For a 2D vector, pointing right and up is positive in the x and y directions, respectively. A vector with a zero x component will point up or down, having only a value in the y direction. Likewise, a vector with a zero y component will only point left and right. All two-dimensional vectors are assumed to have a zero z component, existing only in the x–y plane where $z = 0$.

Trigonometry plays a key role in finding the components of a vector. Since **i** and **j** point in the positive x and y directions, these component vectors form a right triangle. Using simple trigonometric equations and relationships allows one to determine the size of the x and y components for a vector.

Example 13.1 (2D Force Components) Given a 10 N force vector described in the following figure:

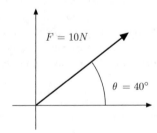

The x and y components of this vector can be found by using simple trigonometry. Realize that this vector is the hypotenuse of a right triangle where the F_x and F_y components are the sides of the triangle:

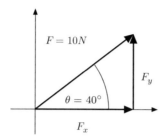

It is obvious that $F = F_x + F_y$. Seeing that F_x and F_y are perpendicular, it is a right triangle. This means that from trigonometry, the following relationships hold:

$$\cos(\theta) = \frac{F_x}{F}$$

$$\sin(\theta) = \frac{F_y}{F}$$

The value for θ is known to be $40°$ and the force magnitude F is $10\,\mathrm{N}$. The relationships become:

$$\cos(40°) = 0.766 = \frac{F_x}{10}$$

$$\sin(40°) = 0.643 = \frac{F_y}{10}$$

leading to values of $F_x = 7.66\,\mathrm{N}$ and $F_y = 6.43\,\mathrm{N}$. This makes sense, because we expect F_x to be greater than F_y since $\theta = 40°$ is $< 45°$.

A vector can be described by the direction and magnitude. The magnitude is a scalar value representing the size of a vector.

Magnitude of a 2D Vector—The magnitude of a force vector expressed as $F = F_x \mathbf{i} + F_y \mathbf{j}$ is found using the formula:

$$|F| = \sqrt{(F_x)^2 + (F_y)^2}$$

Note that from this definition, the magnitude of a vector cannot be negative. However, a vector can point in the opposite direction. This is seen when components of a vector point in the negative x or negative y direction.

Example 13.2 (Force Magnitude) Given the vector F :

$$F = -4\mathbf{i} - 3\mathbf{j}$$

This vector points down and to the left, in the negative x and negative y directions. The magnitude is given by:

$$|F| = \sqrt{(F_x)^2 + (F_y)^2} = \sqrt{(-4)^2 + (-3)^2}$$
$$= \sqrt{16 + 9} = \sqrt{25}$$
$$|F| = 5$$

13.1.2   Vector Addition

Multiple forces can act on an object. The total resultant force is just the sum of the forces acting on the object. This resultant force is found by adding the force vectors. Adding forces is often visualized as adding forces "head to tail".

Adding forces head to tail can become quite difficult to do if the geometry is complex. An easier way to add forces together is to break down the vectors into components and add the individual components. To add two or more 2D force vectors together:

1. Break the vectors into x and y components.
2. Add up the x components, taking care with positive and negative values.
3. Add up the y components, taking care with positive and negative values.
4. Determine the direction of the force vector from the resulting x and y components.

Sum of Forces—The sum of the forces acting on an object is the sum of the individual components, given as:

$$F = (\Sigma F_{ix})\,\mathbf{i} + (\Sigma F_{iy})\,\mathbf{j} = F_x\mathbf{i} + F_y\mathbf{j}$$

where F_{ix} and F_{iy} are the respective magnitudes of force i along the unit vector components $\mathbf{i}$ and $\mathbf{j}$.

Example 13.3 (2D Force Addition) Consider adding two forces: $F_1 + F_2 = F_3$, where F_1 is magnitude 20 at $35°$ and F_2 is magnitude 36 at $165°$. Forces F_1 and F_2 are shown below.

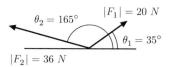

First, the forces must be expressed in terms of their Cartesian component vectors. For force F_1, $F_1 = F_{1x}\mathbf{i} + F_{1y}\mathbf{j}$

$$F_{1x} = |F_1| \cos (\theta_1) = (20) \cos (35°) = 16.38 \,\text{N}$$
$$F_{1y} = |F_1| \sin (\theta_1) = (20) \sin (35°) = 11.47 \,\text{N}$$

Similarly, for F_2, the component vectors are:

$$F_{2x} = |F_2| \cos (\theta_2) = (36) \cos (165°) = -34.77 \,\text{N}$$
$$F_{2y} = |F_2| \sin (\theta_2) = (36) \sin (165°) = 9.31 \,\text{N}$$

Note that F_{2x} is negative. This makes sense because F_2 points to the left. Both F_1 and F_2 are pointing up, so both F_{1y} and F_{2y} are positive. To find the resulting vector, we add the x and y components:

$$F_{3x} = F_{1x} + F_{2x} = 16.38 + (-34.77) = -18.4$$
$$F_{3y} = F_{1y} + F_{2y} = 11.47 + 9.31 = 20.8$$

The vector F_3 points up and to the left since F_{3x} is negative and F_{3y} is positive. Since the components F_{3x} and F_{3y} are similar in magnitude, the angle should be close to $45°$ to the left.

$$\tan (\theta_3) = \frac{F_{3y}}{F_{3x}}$$

$$\theta_3 = \arctan \left(\frac{F_{3y}}{F_{3x}} \right) = \arctan \left(\frac{20.8}{18.4} \right) = 48.5°$$

This is the angle with respect to the negative x axis. To determine the angle from the positive x axis,

$$\theta_3 = 180° - 48.5° = 131.5°$$

The magnitude of F_3 is found as:

$$|F_3| = \sqrt{(F_{3x})^2 + (F_{3y})^2}$$
$$= \sqrt{(-18.4)^2 + (20.8)^2}$$
$$|F_3| = 27.8\,\text{N}$$

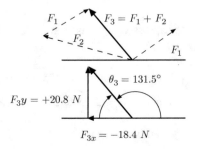

13.2   Moments

A force can induce rotation in an object. The angular rotational force is called a rotational moment. This angular rotational force is also called torque. The moment (torque) on an object depends on the amount of force exerted and the distance from the pivot point.

There are numerous definitions used to describe moments of an object.

Pivot Point—The pivot point is the point an object rotates around. This is usually the center of gravity for the object.

Lever Arm—The lever arm is the distance from the pivot point to where the rotational force is applied.

Line of Action—The line of action is the component of the force inducing rotation which is perpendicular to the lever arm.

> **Sense of Rotation**—The sense of rotation is either ClockWise (CW) or Counter ClockWise (CCW) depending on the direction the force induces rotation.

> **Moment**—The moment induced by a force is the product of lever arm distance d with the magnitude of the force acting along the line of action.
>
> $$M_o = F\,d$$
>
> The force component F is assumed to be positive. A CW rotation is negative, while a CCW rotation is positive.

For more complex 2D systems, it is useful to break a moment calculation into two parts, a horizontal lever arm and a vertical lever arm. The vertical force creates a moment with the horizontal lever arm, and the horizontal force creates a moment with the vertical arm. The two moments add together for the total moment around the pivot point.

Example 13.4 (Moment Calculation) Calculate the moment for the following diagram:

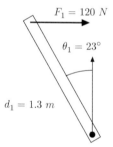

There are two ways to approach this problem. From the definition of a moment, the force component perpendicular to the lever arm can be used to determine the moment. The force in this problem only exerts force in the horizontal direction with a force of 120 N. The force therefore must exert on the lever arm at an angle of 67°. To find the perpendicular force component, the angle between F_1 and $F_{1\perp d_1}$ must be 23° because $F_{1\perp d_1}$ must act on the arm at an angle of 90°. Therefore, the force is determined to be 120N cos (23°) = 110.4 N, and the moment calculation becomes:

$$M = F_{1\perp d_1}\,d_1 = (-)\,110.4\,\text{N}\,(1.3\,\text{m}) = -143.6\,\text{Nm}$$

The negative sign is included because the moment is acting in the clockwise direction.

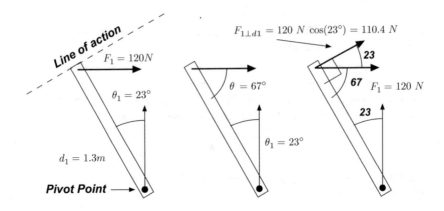

In many cases, it is easier to break a moment into Cartesian components. The lever arm can be split into horizontal and vertical components. The resulting moment in this case is:

$$M = (-)F_1\, d_{1x} = (-)\,120\,\text{N}\,(1.197\,\text{m}) = -143.6\,\text{Nm}$$

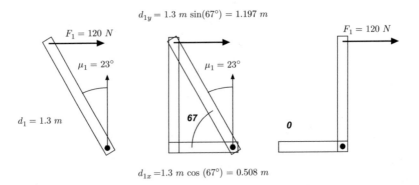

There is no vertical force component acting on the horizontal portion of the lever arm.

Example 13.5 (Complex Moment Calculation) Calculate the total moment for the following diagram:

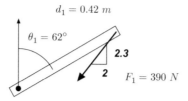

$$d_1 = 0.42 \; m$$

First, the force must be broken into component elements. One way to accomplish this is by using trigonometry. To find the angle of the force with the vertical axis, we know:

$$\tan (\theta_F) = \frac{2}{2.3}$$

$$\theta_F = 41.0°$$

Therefore, the magnitude of the x and y components of the force vector can be determined as:

$$F_{1x} = (390) \sin (41°) = 255.9 \, \text{N}$$

$$F_{1y} = (390) \cos (41°) = 294.3 \, \text{N}$$

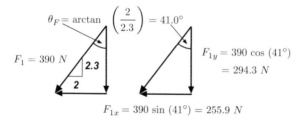

In a similar manner, the lever arm can be broken into horizontal and vertical components as:

$$d_{1x} = d_1 \sin (\theta_1) = 0.362 \, \text{m}$$

$$d_{1y} = d_1 \cos (\theta_1) = 0.191 \, \text{m}$$

The resulting moment is now divided into two separate moments. The vertical component of the lever arm (length d_{1y}) "slides" over to the pivot point. The horizontal component of the force is assumed to work on the vertical arm. Similarly, the vertical component of force works on the horizontal arm.

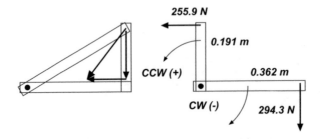

The individual moments are found as:

$$M_1 = (+)(255.9\,\text{N})(0.191\,\text{m}) = \quad 48.88\,\text{(Nm)}$$

$$M_2 = (+)(294.3\,\text{N})(0.362\,\text{m}) = -106.5\,\text{(Nm)}$$

$$M = M_1 + M_2 = -57.6\,\text{(Nm)}$$

Alternatively, the same value can be found "the hard way" by determining the component of the force perpendicular to the lever arm using knowledge about right triangles. Knowing that the lever arm is at $62°$ with the vertical axis, the complement is $28°$ and the angle of the lever arm with the vertical axis must also be $62°$. The angle of the force vector with the lever arm must therefore be $62° - 41° = 21°$. The force vector makes a triangle with the lever arm, and the component perpendicular to the lever arm can be found using the $\sin(21°)$.

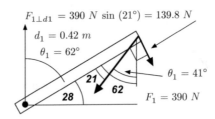

The moment is simply found by multiplying the lever arm length with the force component perpendicular. Since the force is acting in the CW direction, the moment should be negative:

$$M = (-)(0.42\,\text{m})(139.8\,\text{N}) = -58.7\,\text{(Nm)}$$

The two calculated values are off by about 1 Nm due to rounding errors.

13.3 Static Equilibrium

Something in static equilibrium is either totally stationary or moving at a constant velocity (not accelerating). In cases considered here, an object in static equilibrium is not moving. For something to be in static equilibrium, the following must hold true in the 2D case:

1. Forces in the x direction must sum to 0.
2. Forces in the y direction must sum to 0.
3. Moments must sum to 0

Example 13.6 (Static Equilibrium) Given the following free body diagram, solve for the unknown values F_1, F_2, and d_2 assuming static equilibrium for the system.

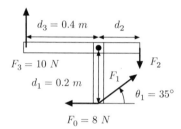

The sum of forces in the x direction can be written as:

$$-F_0 + F_1 \cos (35°) = 0$$

The sum of forces in the y direction can be written as:

$$F_1 \sin (35°) - F_2 + F_3 = 0$$

The sum of moments equation can be written as:

$$(-)F_0 d_1 + (+)F_1 \cos (35°) d_1 + (-)F_2 d_2 + (-)F_3 d_3 = 0$$

Substituting known values, the equations become:

$$-8 + F_1 \cos (35°) = 0$$
$$F_1 \sin (35°) - F_2 + F_3 = 0$$
$$-8(0.2) + F_1 \cos (35°) (0.2) - F_2 d_2 - 10(0.4) = 0$$

Using the first equation, F_1 is found to be 9.77 N. Using the second equation, F_2 is found to be 15.6 N. Using the third equation, d_2 is found to be −0.256 m.

Determining a negative value for a lever arm or a force magnitude is not wrong. This means that the free body assumption in the drawing was incorrect. Finding a negative d_2 value means that the lever arm should have been drawn to the left of the pivot point instead of the right.

Redrawing the free body diagram with d_2 to the left and changing the sign of the d_2 moment in the third equation result in the following diagram and static equilibrium equations:

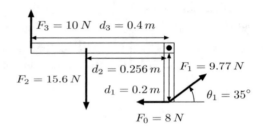

$$-8 + (9.77) \cos (35°) = 0$$
$$(9.77) \sin (35°) - (15.6) + 10 = 0$$
$$-8(0.2) + (9.77) \cos (35°) (0.2) + (15.6) (0.2564) - 10(0.4) = 0$$

Problems

13.1 For the following figure:

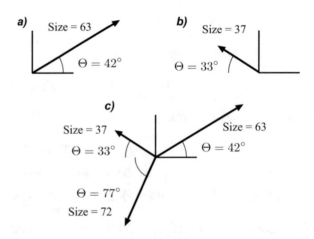

(a) Determine the x and y components of vector F_a.
(b) Determine the x and y components of vector F_b.
(c) Determine the sum of the three vectors. For the resultant vector, find its magnitude and angle (in degrees) with the positive x axis.

13.2 For the following figure:

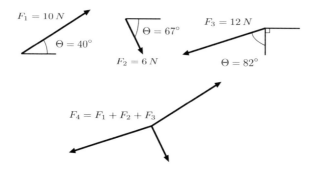

(a) Determine the x and y components of vector F_1
(b) Determine the x and y components of vector F_2
(c) Determine the x and y components of vector F_3
(d) Determine the x and y magnitude of vector F_4 and the sum of F_1, F_2, and F_3.
(e) Determine the angle of vector F_4 with respect to the positive x axis.

13.3 Calculate the moments for the following cases:

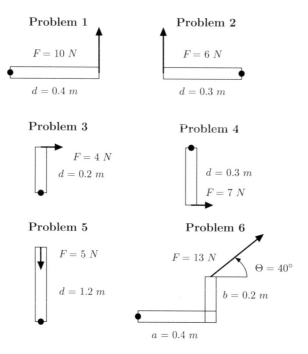

13.4 Calculate the moments for the following cases:

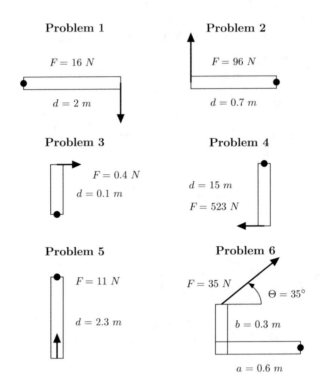

13.5 Write the equations of static equilibrium for the following free body diagram:

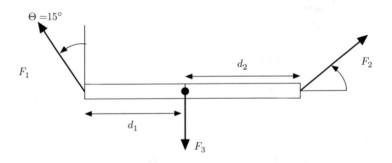

For the following free body diagram:

Write the equations of static equilibrium in terms of variables (d_1, d_2, m, θ, F_1, etc.).

Set up the problem in the form of $Ax = b$ if linear.
Solve for the unknown values if possible.

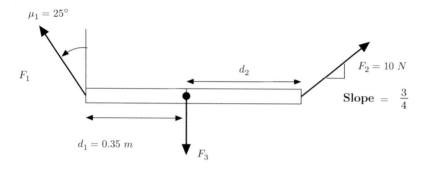

13.6 For the following free body diagram:
Write the equations of static equilibrium in terms of variables (d_1, d_2, m, θ, F_1, etc.).
Set up the problem in the form of $Ax = b$ if linear.
Solve for the unknown values if possible.

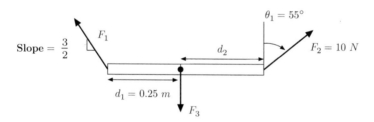

13.7 For the following free body diagram:
Write the equations of static equilibrium in terms of variables (d_1, d_2, m, θ, F_1, etc.).
Set up the problem in the form of $Ax = b$ if linear.
Solve for the unknown values if possible.

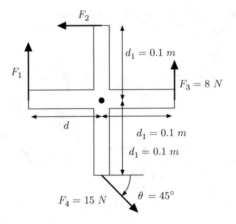

13.8 For the following free body diagram:

Write the equations of static equilibrium in terms of variables (d_1, d_2, m, θ, F_1, etc.).

Set up the problem in the form of $Ax = b$ if linear.

Solve for the unknown values if possible.

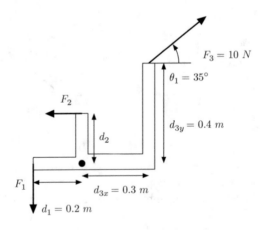

Chapter 14
Differential Equations

14.1 Ordinary Differential Equations

Ordinary Differential Equations only depend on one dependent variable. In many cases, the dependent variable is time. In some stead-state cases, the dependent variable may be a dimension like position x along an axis or radial position r in a cylinder.

14.1.1 Dynamic Mass Balance Example

Consider a simple tank of constant cross-sectional area A holding a liquid. The height of the tank, $h(t)$, varies with time. The volumetric flow rate into the tank may also vary with time, $F_{in}(t)$. The flow rate out from the bottom of the tank is assumed to be proportional to the square root of the tank height, $F_{out}(t) = k\sqrt{h(t)}$. The liquid density ρ is assumed constant, as is the molecular weight of the liquid, M. The tank area A is also assumed constant. The total number of moles of a liquid in the tank is given by:

$$\frac{1}{M}\rho V$$

where M is the molecular weight in g/mol, ρ is the density in g/L, and V is the volume in L. Since volume is known to change with time and the cross-sectional area is constant, this can be written as:

$$\frac{1}{M}\rho V(t) = \frac{1}{M}\rho A h(t)$$

© Springer Nature Switzerland AG 2022
E. Gatzke, *Introduction to Modeling and Numerical Methods for Biomedical and Chemical Engineers*, https://doi.org/10.1007/978-3-030-76449-4_14

Since the tank area and liquid density are constant, the accumulation rate for the liquid in the tank is:

$$\frac{1}{M}\rho A \frac{dh}{dt}(t)$$

The overall balance equation may be written as:

$$\frac{1}{M}\rho A \frac{dh}{dt}(t) = \frac{1}{M}\rho F_{in}(t) - \frac{1}{M}\rho F_{out}(t)$$

Here, each term has units of mol/time. This makes sense, as all terms are rate terms: accumulation rate, rate in, and rate out.

Canceling the density and molecular weight terms and substituting the expression for $F_{out}(t)$ leads to:

$$A\frac{dh}{dt}(t) = F_{in}(t) - k\sqrt{h(t)}$$

The solution to this differential equation is not a single scalar value. Rather, the solution will determine the value for $h(t)$, a function of time. The solution must satisfy the governing mass balance equation at all points in time.

Consider a slightly different linear dynamic case. For the same simple tank system, one may assume that $F_{in}(t)$ is zero and the flow rate from the tank is a linear form $kh(t)$. The mass balance becomes:

$$A\frac{dh}{dt}(t) = -kh(t)$$

Given that the height in the tank at time $t = 0$ is h_o, the solution of this linear differential equation is:

$$h(t) = h_o e^{-\frac{k}{A}t}$$

To verify this, take the derivative of $h(t)$:

$$\frac{dh}{dt}(t) = -\frac{kh_o}{A}e^{-\frac{k}{A}t}$$

Substitute both $h(t)$ and $\frac{dh}{dt}(t)$ into the original equation:

$$A\left(-\frac{kh_o}{A}e^{-\frac{k}{A}t}\right) = -k\left(h_o e^{-\frac{k}{A}t}\right)$$

From this, it appears that the function suggested for $h(t)$ satisfies the differential equation. However, most real engineering situations are nonlinear and often the forcing functions (like $F_{in}(t)$) are very complex. In these complicated cases, numerical methods can be used to approximate the solution. In this simple linear case, the value for h can be determined at any point in time by evaluating the analytical expression. When using numerical methods, the value for $h(t)$ is typically approximated at discrete time values.

14.2   Euler Integration for ODEs

Consider a general differential equation in the form:

$$\frac{dy}{dt}(t) = f(t)$$

The goal is to find an approximate value for $y(t)$ given that only information concerning the rate of change $\frac{dy}{dt}$ is known. In many engineering cases, the initial conditions are known. This means that the value of y at time $t = 0$ is given. The exact derivative can be approximated as:

$$\frac{dy}{dt}(t) \approx \frac{y(t + \Delta t) - y(t)}{\Delta t}$$

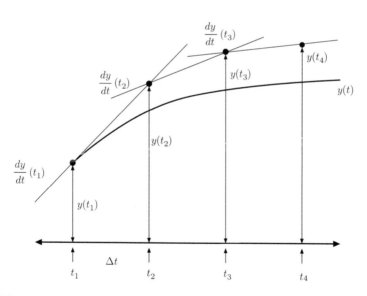

Fig. 14.1 Euler integration to approximate the solution to a simple differential equation given by $\frac{dy}{dt}(t) = f(t)$

leading to:

$$\frac{y(t + \Delta t) - y(t)}{\Delta t} \approx f(t)$$

This can be rearranged as:

$$y(t + \Delta t) \approx y(t) + \Delta t \; f(t)$$

This equation means that given the value for $y(t)$ and $f(t)$ at the current time t, the value for $y(t + \Delta t)$ can be approximately calculated. The approximation becomes more accurate for smaller values of Δt.

Note that there are different derivative approximations that are available. Some are more accurate but more complex. In this initial discussion, a simple forward difference is presented. Runge–Kutta methods are commonly used for ODE interaction because they offer much higher accuracy than basic forward difference Euler integration. However, Runge–Kutta methods require evaluating the function $f(t)$ multiple times.

Example 14.1 (Euler Integration) Consider the differential equation for water draining from a tank:

$$A\frac{dh}{dt}(t) = 0 - k\sqrt{h(t)}$$

With values of $A = 2$ and $k = 18$, this can be rearranged to the form:

$$\frac{dy}{dt}(t) = -9\sqrt{y(t)}$$

where $h(t) = y(t)$. The initial tank height must be given and the step size must be fixed.

Assuming that $y(t = 0) = 10$ and $\Delta t = 0.1$, you are to find future values of $y(t)$ using Euler integration. The general equation is:

$$y(t + \Delta t) \approx y(t) + \Delta t \; f(t)$$

This will be used with the assumption that the right hand side (RHS) function is given by $f(t) = -9\sqrt{y(t)}$. This means that the update equation is:

$$y(t + \Delta t) \approx y(t) + \Delta t \left(-9\sqrt{y(t)}\right)$$

Given that $y(t = 0) = 10$, one may find $y(t = 0.1)$ using the equation above:

$$y(t = 0.1) = y(t = 0) + \Delta t \; f(t = 0)$$

$$= 10 + (0.1) \left(-9\sqrt{10}\right) = 7.15$$

The value for $y(t = 0.1)$ can be used to find $f(t = 0.1)$, the RHS of the differential equation. Together, these two values can then be used to find $y(t = 0.2)$ as:

$$y(t = 0.2) = y(t = 0.1) + \Delta t \ \ f(t = 0.1)$$

$$= 7.15 + \Delta t \ \left(-9\sqrt{7.15}\right) = 4.75$$

The rest of the values are computed as follows:

Iteration	t	$y(t)$	$f(t)$	$\Delta t \ f(t)$
1	0	10.0	−28.5	−2.85
2	0.1	7.15	−24.1	−2.41
3	0.2	4.75	−19.6	−1.96
4	0.3	2.79	−15.0	−1.50
5	0.4	1.29	−10.2	−1.02
6	0.5	0.270	−4.68	−0.468

Interestingly, the value for $y(t = 0.6)$ is:

$$y(t = 0.6) = 0.270 + \Delta t \ \left(-9\sqrt{0.270}\right) = -0.198$$

A negative value will lead to imaginary numbers in the next iteration. This means that the accuracy of the method on this problem with the given value for Δt is probably quite poor. A smaller step size is required to accurately approximate this solution. This equation represents a real tank system that cannot have a negative level, so the approximation is inaccurate and should be calculated using a much smaller value of Δt.

Euler integration is a simple way to approximate Ordinary Differential Equations (ODEs) given that an initial value for the differential equation is known. As seen in the previous example, the approximation may lead to significant errors. Effectively, the value for the function is known at the initial time t_1. Given the value for $y(t_1)$, the rate of change is calculated: $\frac{dy}{dt}(t_1)$. A line can be constructed from the value $y(t_1)$, given that the slope of the line is $\frac{dy}{dt}(t_1)$. Using this line approximation allows one to determine the value at time $t_2 = t_1 + \Delta t$, leading to $y(t_2)$. This is illustrated in Fig. 14.1. Note again that smaller values for Δt will lead to a better approximation for the solution of the true function, $y(t)$. However, this will require more calculations to be performed.

14.3   Boundary Value ODE Problems

Given two first-order coupled ODEs, the initial conditions are typically specified.
The results can be determined using Euler integration. A second-order differential
equation can be converted into two first-order ODEs. Consider a one-dimensional
ballistics problem:

$$F = ma = m\frac{dv}{dt} = m\frac{d^2x}{dt^2}$$

This second-order ODE $F = md^2x/dt^2$ requires two initial conditions. This
equation can be written as two first-order ODEs:

$$F = m\frac{dv}{dt}$$

$$v = \frac{dx}{dt}$$

Knowing the initial position $x(0)$ and initial velocity $v(0)$, adequate initial conditions are provided for integration using Euler's method.

Now, consider a case where instead of two initial conditions, you specify the
initial position $x(0)$ and the final position $x(t_f)$. This means you know the starting
position and the position at some future time t_f but the initial velocity is not
specified.

To approach this problem, one could specify the initial velocity $v(0)$ and then
integrate the two equations from $t = 0$ to $t = t_f$. If the final position $\hat{x}(t_f)$
matches the specified condition $x(t_f)$, the guess was perfect and you have solved
the problem.

However, if the initial guess was not perfect, you must iterate. This can be seen
as searching for the initial velocity $v(0)$ to satisfy the following equation:

$$f(v(0)) = x(t_f) - \hat{x}(t_f) = 0$$

There is only one unknown value, $v(0)$. This appears to be a single algebraic equation that could be solved by Newton's method or bisection. However, determination
of the final value $\hat{x}(t_f)$ requires integrating the differential equations from $t = 0$ to
$t = t_f$. There are actually many unknown values for x and v, which all depend on
the specified value of the initial velocity $v(0)$.

Problems requiring shooting method approaches appear in a variety of domains.
For one-dimensional steady-state heat transfer, the independent value may be
positioned in the material x. The temperature at two boundaries may be specified,
$T(0)$ and $T(x = L)$. Similarly, in steady-state one-dimensional fluid flow, the fluid
velocity may be specified at two boundaries rather than specifying both the velocity
and rate of change in velocity at one of the boundaries.

Iteration	t	$x(t)$	$v(t)$
1	0	$x(0)$	???
2	0.1	$x(0.1)=f_{11}(x(0),v(0))$	$v(0.1)=f_{12}(x(0),v(0))$
3	0.2	$x(0.2)=f_{11}(x(0.1),v(0.1))$	$v(0.2)=f_{12}(x(0.1),v(0.1))$
$\vdots$	$\vdots$	$\vdots$	$\vdots$
N	t_f	$x(t_f)$	$v(t_f)$

14.4 Partial Differential Equations

Partial Differential Equations occur when the dependent variable depends on more than one independent variable. For many engineering problems, the variable of interest may be temperature, concentration, fluid velocity, or internal stress. The dependent variables are usually positioned in the x, y, or z direction. Time is an independent variable for dynamic problems.

14.4.1 Heat Transfer in Solids

Given a design for a single piece of equipment, the temperature can be determined through simulation. For a steady-state problem, the temperature of the object will be a function of x and y in 2D simulations and x, y, and z for 3D models. The thermal conductivity and heat capacity of the material must be specified as physical parameters. Boundary conditions must be described. A surface could be held at a constant temperature, or a surface could be assumed to be insulating (no heat transferred across that boundary). A solid object heated or cooled by a liquid or gas can be modeled with convective heat flux at the boundary. Complex objects can accurately be modeled in this way to give predictions on temperatures inside the solid object. The temperature difference between two points in the solid drives energy to transfer from one point to another.

14.4.2 Fluid Flow

Given a flow path such as a pipe or vein, liquid (or gas) can flow through the system. The viscosity of the liquid determines the resulting velocity profile. In most situations, "no slip" is assumed such that the velocity of the fluid at the wall is 0. Given boundary conditions for the inlet and outlet, COMSOL can determine the resulting values for the liquid velocity as a function of x, y and sometimes z and t. The geometry of interest is just the area where the fluid is flowing; the solid surrounding the liquid need not be modeled in many situations. The pressure difference between two points in the fluid provides the driving force for fluid flow.

14.4.3   Reaction and Diffusion of Dilute Species

In many cases, the aqueous concentration may be modeled for a system. Consider a system where a low-concentration drug is being released into a system. The concentration difference between two points in the liquid drives diffusion of the species. A location with a high concentration will "push" the dilute species into an area with lower concentration. The rate of diffusion depends on a diffusion coefficient. If there is fluid flow, the dilute species can also be transferred as it is carried along with the liquid. In other cases, a chemical reaction may occur at a boundary where the drug is "consumed" as it reacts or crosses the volume edge.

14.4.4   Stress–Strain for Elastic Materials

In mechanical systems, the elastic material properties can be used to evaluate a design for a single piece of equipment. Consider designing a mechanical device. The geometry of the device can be used to describe the item. The Young's modulus for the material determines the stiffness of the material. In simulation, one or more of the device boundaries can be assumed "fixed," while a force can be applied at a location or along a boundary. The simulation will then determine internal stress values for the simulation.

14.5   Finite Difference Approximation for PDEs

To find numerical solutions to a PDE problem, the domain of interest can be discretized. Similar to Euler integration, expressions for the approximate value of the variable of interest are used to define the equations to be solved. The independent variable is discretized into segments of size h. The forward difference for a variable $f(x)$ is given by:

$$f'\Big|_x = \frac{f(x+h) - f(x)}{h}$$

while the backward difference is given by:

$$f'\Big|_x = \frac{f(x) - f(x-h)}{h}$$

The second derivative is the rate of change for the first derivative. One approximation for the second derivative is given as:

$$f''\Big|_x = \frac{\left(\frac{f(x+h)-f(x)}{h}\right) - \left(\frac{f(x)-f(x-h)}{h}\right)}{h}$$

$$= \frac{f(x+h) - 2f(x) + f(x-h)}{h^2}$$

There are many other types of approximations that may be used to approximate the derivatives specified in the governing differential equation.

14.5.1 PDE Boundary Conditions

To solve a PDE, one must specify appropriate boundary conditions. Some common types of boundary conditions used in PDE solution include:

Dirichlet Conditions are fixed or constant at the boundaries. Example: holding the temperature or velocity at the domain edge to one value; the constant concentration in a diffusion problem; fluid moving between two flat plates with fixed velocity.

Neumann A derivative is specified at the boundary. Example: the heat flux or reaction rate on the edge of the domain is specified.

Mixed Boundary conditions are not required to all be the same. For example, in heat transfer problems, some boundaries may be assumed to be held at a constant temperature, while other boundaries could have the heat flux specified. For a well-insulated boundary, the flux would be set to 0. Some boundaries may be set to follow Newton's law for convective cooling if the boundary is exposed to a gas or liquid at a specified temperature.

Initial For time-dependent problems, the initial conditions are typically specified across the whole domain. For a dynamic heat transfer problem, the temperature for all points at time $t = 0$ can be specified, along with the boundaries of the domain.

14.6 Simple PDE Example

Consider a 2D steady-state conduction problem. Imagine a rectangular piece of metal extending vertically, with each of the four sides held at a different temperature. The problem is assumed to be at steady-state so that the temperature does not depend on time. The vertical dimension is assumed so large that any horizontal slice would be representative of the temperature such that temperature does not depend on elevation z. This will not be true at the very top and bottom of the metal where end effects may change the result. However, to simplify the problem, temperature inside the material can be assumed only as a function of only x and y. The following governing equation to find $T(x, y)$ may be used:

$$0 = \alpha \left(\frac{\partial^2 T}{\partial x^2} + \frac{\partial^2 T}{\partial y^2} \right)$$

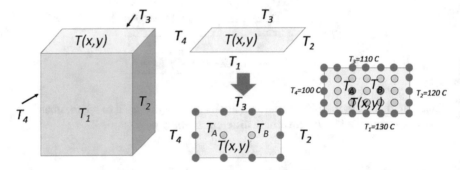

Fig. 14.2 2D steady-state conduction for a solid with 2:3 aspect ratio. The accuracy improves if smaller discretization is applied, resulting in more equations to be solved

To simplify discretization, the domain will be divided into equal area squares with each side of length h, so that $\Delta x = \Delta y = h$. A very coarse discretization can be constructed for as shown in Fig. 14.2, where the aspect ratio of the material is 2:3. Assuming that each of the four boundaries is held constant at temperatures T_1, T_2, T_3, and T_4, there are now only two unknown temperature values: T_A and T_B. Using a smaller interval for h will result in a more accurate estimate; however, there will be many more equations as a result.

For the simple case, two equations may be written based on the governing equation. We know that the second-order finite difference can be written as:

$$f''\Big|_x = \frac{f(x+h) - 2f(x) + f(x-h)}{h^2}$$

Now, the problem is examining temperature T as a function of x and y so that this expression can now be written as:

$$\frac{\partial^2 T}{\partial x^2} = \frac{T(x+\Delta x) - 2T(x) + T(x-\Delta x)}{\Delta x^2}$$

$$\frac{\partial^2 T}{\partial y^2} = \frac{T(y+\Delta y) - 2T(y) + T(y-\Delta y)}{\Delta y^2}$$

This means that the temperature T at any point x, y will depend on the value of points left, right, above, and below the point at x, y. Using the governing equation at point T_A,

$$0 = \alpha \left(\left(\frac{T(x+\Delta x) - 2T(x) + T(x-\Delta x)}{\Delta x^2} \right) + \left(\frac{T(y+\Delta y) - 2T(y) + T(y-\Delta y)}{\Delta y^2} \right) \right)$$

This simplifies to:

$$(T_B - 2T_A + T_4) + (T_3 - 2T_A + T_1) = 0$$

or simply:

$$-4T_A + T_B = -T_1 - T_3 - T_4$$

Similarly, for T_B, the resulting governing equation is:

$$(T_A - 2T_B + T_2) + (T_3 - 2T_B + T_1) = 0$$

or simply:

$$T_A - 4T_B = -T_1 - T_2 - T_3$$

The boundary temperatures T_1, T_2, T_3, and T_4 are specified. The only two unknowns are T_A and T_B. The set of equations is linear and can be written in the form of $Ax = b$ as:

$$\begin{bmatrix} -4 & 1 \\ 1 & -4 \end{bmatrix} \begin{bmatrix} T_A \\ T_B \end{bmatrix} = \begin{bmatrix} -T_1 - T_3 - T_4 \\ -T_1 - T_2 - T_3 \end{bmatrix}$$

Soving this problem with $T_1 = 110$ C, $T_2 = 120$ C, $T_3 = 130$ C, and $T_4 = 100$ C results in $T_A = 114.66$ C and $T_B = 118.66$ C. Increasing the resolution of the discretization results in more linear equations as shown in Fig. 14.2. The solution of these problems is shown in Fig. 14.3 as the size of $h = \Delta x = \Delta y$ decreases. The values for T_A and T_B converge to what is assumed the true solution.

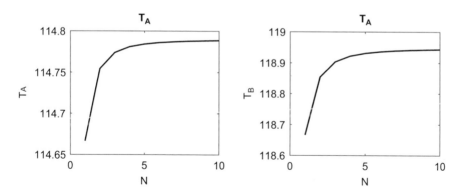

Fig. 14.3 Convergence of the 2D steady-state conduction problem for T_A and T_B as the discretization resolution is increased

14.7   Introduction to COMSOL

COMSOL is a powerful simulation environment. It allows users to describe fairly complex geometric shapes using an interface similar to many computer aided drawing packages. A shape can be given physical properties like heat capacity and Young's modulus determined from real materials. Simulation studies can be computed, assuming either steady-state or dynamic behavior. The results can be plotted easily for analysis.

While COMSOL can do many different things, a few types of simulations are often first considered at an introductory level. While many simulations are best modeled in 3D, it is a good idea to start by working with a 2D approximation. Also, many simulations are best simulated as time-dependent. However, it is often simpler to approach problems with a steady-state model if possible. If possible, reduce the dimensionality of the problem as much as possible, within reason.

To simulate a system, COMSOL must know the geometry, the physical parameters, and the boundary conditions. Given this information, COMSOL will automatically generate hundreds (or thousands or hundreds of thousands) of equation that must be solved. This is accomplished by discretizing the simulation domain geometry into many mesh points. The value of the variables at each mesh point depends on the values of the neighbor mesh points in a steady-state simulation. Note that COMSOL uses irregular mesh points rather than constant discretization of the domain of the variable. This allows COMSOL to accurately simulate complex non-regular geometries. Adaptive meshing allows for higher resolution simulation in regions that are commonly difficult to simulate due to high gradients, such as corners of a material.

As a result, COMSOL simulations are often limited by computational ability. A 3D simulation requires many more equations than a 2D simulation. A time-dependent dynamic simulation typically will require more equations than a steady-state. Having multiple interacting types of physics giving different governing equations results in more equations.

The simulation results also depend on the physical parameters and material properties. In some cases, these values are not known with significant accuracy. In other cases, these values may not be constant values. Determining how the material properties change can be quite difficult. For example, the viscosity of a fluid may depend on the temperature of the liquid. If the liquid temperature changes in the simulation, the viscosity change may not be considered and the results may not be very accurate.

COMSOL can simulate multiple types of physics. Typically, simulations are easier to accomplish when only a single governing equation considered. COMSOL may be able to simulate blood flow in a vein AND the forces and motion of the wall, and however it may be quite difficult to correctly tie two different domains together in one simulation.

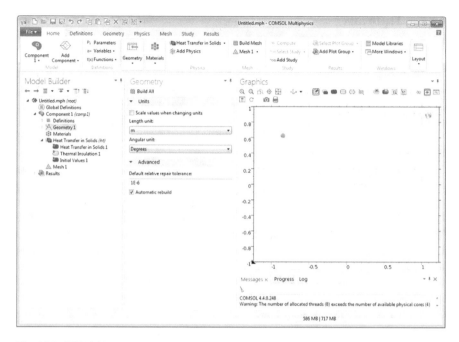

Fig. 14.4 COMSOL interface

COMSOL requires the user to make many assumptions. One must define boundary conditions and initial conditions. In many cases, these assumptions are not very realistic. However, these assumptions must be made such that even an initial simulation calculation may be performed. It is the job of the engineer to have some insight into what parts of the simulation can be simplified or approximated and what must be considered in detail.

COMSOL can simulate multiple types of physics simultaneously. In one tutorial, a fluid is flowing past a heated wire where a reaction takes place. The fluid flow of the liquid is determined by the inlet flow rate and boundary conditions. The flow carries a dilute species. The dilute species may diffuse through the liquid, moving from regions of high concentration to regions of low concentration. The dilute species may also react if it passes by the surface of the heated wire, requiring a reaction rate expression. The temperature of the liquid as it flows and reacts may also be determined. The temperature can affect fluid properties such as density and viscosity. These three types of physics can all be modeled simultaneously; however, it may be very difficult to create such a simulation without following the detailed tutorial. Initial attempts not following along the instructions from a tutorial should try to keep the system model as simple as possible (Fig. 14.4).

14.8 Tutorials

There are numerous tutorials available in COMSOL. Tutorials can be found in the COMSOL help browser. However, the COMSOL tutorial interface may be confusing. In some COMSOL versions, every time you click on the COMSOL interface while in the tutorial, it takes you to a tutorial on that function. You may want to download the PDF tutorial file for any tutorial you want to work through, viewing the PDF file in a separate application while following the instructions in COMSOL.

Remember, you must follow the instructions precisely. Any slight mistake or omission can cause the simulation to fail. Note that COMSOL has a fairly confusing interface, so locating the correct action for each step in the tutorial can be tricky. Examine Fig. 14.4. The ribbon across the top includes a variety of functions. The "Model Builder" area to the left has an outline for the simulation where sections can be expanded or collapsed. Some functions will require use of a third input area in the middle of the screen to the left of the graphics area. The interface is confusing at first, but the basic process is similar for most problems: decide on the basic type of simulation, set up the geometry of the domain of interest, apply physical properties to the domain, set the boundary conditions, run the simulation, and plot/analyze the results.

Six simple tutorials are suggested here. There are many available, but these tutorials provide a basic scope for COMSOL capabilities. These tutorials each take 10-40 min to complete. Other tutorials may be quite complex, requiring hours to fully reproduce. You should take time to peruse the various tutorials that are available.

To open tutorials, click on the "?" in the upper right corner of the application (or hit F1). This should open the Help window on the bottom right portion of the window. Click the button with three horizontal lines that is to the left of the help search box. This should open the outline for COMSOL Help. Scroll to the bottom and expand the last entry, "Application Libraries" then expand "COMSOL Multiphysics". Expanding "Heat Transfer" should indicate options for multiple tutorials, including "Steady-State 2D Heat Transfer with Conduction" and "Asymmetric Heat Transfer". After selecting a tutorial, it is suggested that you select the red PDF button and download the tutorial PDF file or open the PDF file in a new window separate from the COMSOL application.

Steady-State 2D Heat Transfer with Conduction – Consider a long rectangular metal rod where temperature is the dependent variable. Taking a slice of the rod results in a 2D problem that is representative of the overall material under the assumption that end effects are negligible. Each side of the rod can be simulated with different boundary conditions. In this tutorial, some sides are simulated with fixed temperatures and some are assumed to have a fixed energy flux across the boundary. The physical properties of the material can also be varied to examine the effects on the temperature

Asymmetric Heat Transfer – Consider a cylindrical metal material. A cylinder is a 3D object, however it is symmetric across the center line of the cylinder. Rather than simulate the more complicated 3D model, a 2D asymmetric model can be used. The 2D temperature distribution is rotated around the center line. In this tutorial, the cylinder undergoes a transient change at time $t = 0$, changing the boundary values from 0 C to 1000 C.

COMSOL can simulate material deformation under load conditions. By investigating the internal stresses of a material, one may predict if a design should be adequate for the given simulation. Again, click the button with three horizontal lines that is to the left of the help search box. This should open the outline for COMSOL Help. Scroll to the bottom and expand the last entry, "Application Libraries" then expand "COMSOL Multiphysics". Expanding "Structural Mechanics" should indicate options for multiple tutorials, including "Tapered Cantilever with Two Load Cases" and "Stresses and Strains in a Wrench".

Tapered Cantilever with Two Load Cases – COMSOL can simulate internal stresses for a material with complex geometry. This simulation fixes one boundary and applies a force on a separate boundary. The stress and deformation of the material can be determined, given the geometry of the solid and the physical parameters for the material.

Stresses and Strains in a Wrench – COMSOL allows for definition of relatively simple geometries for the domain of interest. More complex geometries can be obtained in COMSOL using Boolean operations. However, some simulations may include very complex domain geometries developed in other programs. This tutorial loads a complex 3D model and simulates the stress and deformation for the object under a specified load.

COMSOL can simulate much more complex systems. Moving from steady-state to dynamic simulation can significantly increase the complexity of a simulation by adding time as an independent variable. Additionally, fluid flow presents interesting mathematical and numerical challenges. Again, click the button with three horizontal lines that is to the left of the help search box. This should open the outline for COMSOL Help. Scroll to the bottom and expand the last entry, "Application Libraries" then expand "COMSOL Multiphysics". Expanding "Fluid Dynamics" should indicate options for multiple tutorials, including "Flow Past a Cylinder".

Flow Past a Cylinder – Simulation of problems involving fluid dynamics can be especially difficult. Under certain conditions, fluids will exhibit turbu- lent flow with complex eddies and vortices. The simulation of turbulent flow may not exactly reproduce the real-world but it can be used to give guidance on what type of flow to expect. This tutorial includes a 2D geometry for flow between two flat plates with a single rod obstructing the flow. The flow is assumed to be dynamic, with no initial fluid velocity.

Simulating multiple dependent variables which affect one another can in- crease simulation complexity. Click the button with three horizontal lines that is to the left of the help search box. This should open the outline for COMSOL Help. Scroll to the bottom and expand the last entry, "Application Libraries" then expand

"Chemical Reaction Engineering Module" instead of "COMSOL Multiphysics". Expanding "Reactors with Mass and Heat Transfer" should indicate options for multiple tutorials, including "Thermal Decomposition".

Thermal Decomposition – This steady-state model simulates flow in a flat reactor system with a non-rectangular irregular geometry. A heated rod crosses the reactor. Fluid velocity, temperature, and reactant concentration are all simulated in the 2D domain. Material properties such as density and viscosity can be simulated as temperature-dependent. Reaction rate expressions can be simulated as a nonlinear function. This tutorial demonstrates that multiple types of physics can be used to simulate interactions between momentum transfer, heat transfer, and mass transfer.

14.9   COMSOL Design

When developing a simulation model for a system, consider the following:

- Try to simulate the process in 2D instead of 3D

    - A 2D cross section may be accurate enough to describe the process.
    - Example: discretizing each dimension into 1,000 points results in 1,000,000 equations for the 2D case and 1,000,000,000 for the 3D case.

- Only consider one domain

    - Pick one material domain of interest.
    - Temperature, velocity, and concentration may all affect one another. However, simulating multiple types of physics can be challenging for experienced engineers.
    - There may be two separate materials with differing physical properties. Again, correctly simulating this type of multi-domain problem can often be problematic.

- Try to consider a steady-state case

    - Dynamic systems are more difficult to simulate and analyze.
    - The real system may be dynamic, but specifying appropriate boundary conditions and initial conditions can often be quite challenging.
    - Additionally, adding another dimension such as time is similar to moving from 2D to 3D simulation in computational complexity.

- Simplify the geometry as much as possible

    - A detailed geometric model may not be needed.
    - An approximate cross section may be adequate for giving some insight into the problem.
    - Adding more geometric complexity may just increase the computational requirements without significantly increasing simulation accuracy.

- Build up from something that works

  - Get a very simple model that runs and then add more complexity.
  - Iteratively refining the simulation may be a better approach rather than jumping into large amounts of complexity initially.

14.9.1 Parametric Study

A simulation is created to provide useful information. However, COMSOL simulations can provide an overwhelming amount of information in a single simulation. As a result, a parametric study may be performed. First, the variable of interest must be determined.

> **Variable of Interest**—A single scalar value determined in a simulation. It typically changes when a simulation parameter changes.

Since the variable of interest is a scalar value, the simulation results must be analyzed to determine a single value. Some examples include:

- The temperature or pressure at a single point in the simulation
- The average exit concentration
- The maximum or minimum stress in a simulation
- The concentration value at a point in time.

The variable of interest should change when simulation parameters change.

> **Model Parameters**—A value in a simulation that is assumed to be adjustable.

Some examples of model parameters include:

- The conductivity of a heated material
- The viscosity of a fluid
- The size of a geometric element (like a blockage)
- The diffusion coefficient for a dilute species
- Young's modulus.

The parametric sensitivity of a model can be approximated by using a simulation model to evaluate the effect of different values of model parameters in order to determine the effect on a variable of interest. Graphically, this is accomplished by creating graphs of the variable of interest as a function of one or more model parameters. Excel can be used to create these graphs. As a result, COMSOL model simulations must be run repeatedly.

Problems

14.1 Given the equation:

$$\frac{dx}{dt} = -0.2\,x(t)$$

use Euler's integration to integrate the function from $t = 0$ to $t = 5$ using $\Delta t = 1$ as your step and $x(t = 0) = 5$ as the initial value.

14.2 Given the equation:

$$\frac{dx}{dt} = -0.3\,x(t)$$

use Euler's integration to integrate the function from $t = 0$ to $t = 10$ using $\Delta t = 2$ as your step and $x(t = 0) = 25$ as the initial value.

14.3 Given the equation:

$$\frac{dx}{dt} = -0.02\,(x(t))^2$$

use Euler's integration to integrate the function from $t = 0$ to $t = 8$ using $\Delta t = 1$ as your step and $x(t = 0) = 5$ as the initial value.

14.4 Given the equation:

$$\frac{dx}{dt} = 0.1\,(x(t))^2$$

use Euler's integration to integrate the function from $t = 0$ to $t = 3$ using $\Delta t = 0.5$ as your step and $x(t = 0) = 2$ as the initial value.

14.5 Given the equation:

$$\frac{dx}{dt} = -20\,\sqrt{x(t)}$$

use Euler's integration to integrate the function from $t = 0$ to $t = 1$ using $\Delta t = 0.1$ as your step and $x(t = 0) = 100$ as the initial value. What happens at $t = 0.9$? Why is that a problem? What would you suggest to fix this issue?

Further Reading

1. Chapra, S., & Canale, R. (2020). *Numerical methods for engineers* (8th ed.) McGraw-Hill Education.
2. Davis, R.A. (2021). *Practical numerical methods for chemical engineers: Using Excel with VBA* (5th ed.). CreateSpace Independent Publishing Platform.
3. Law, V.J. (2013). *Numerical methods for chemical engineers using Excel, VBA, and MATLAB* (1st ed.). CRC Press.
4. Rao, S.S. (2001). *Applied numerical methods for engineers and scientists* (1st ed.). Pearson.

Index

© Springer Nature Switzerland AG 2022
E. Gatzke, *Introduction to Modeling and Numerical Methods for Biomedical and Chemical Engineers*, https://doi.org/10.1007/978-3-030-76449-4

Printed in the United States
by Baker & Taylor Publisher Services